Salesforce End-to-End Implementation Handbook

A practitioner's guide for setting up programs and projects to deliver superior business outcomes

Kristian Margaryan Jørgensen

BIRMINGHAM—MUMBAI

Salesforce End-to-End Implementation Handbook

Copyright © 2023 Packt Publishing

Group Product Manager: Alok Dhuri
Publishing Product Manager: Akshay Dani
Senior Editor: Rohit Singh
Technical Editor: Maran Fernandes
Copy Editor: Safis Editing
Project Coordinator: Prajakta Naik
Proofreader: Safis Editing
Indexer: Sejal Dsilva
Production Designer: Ponraj Dhandapani
Developer Relations Marketing Executive: Deepak Kumar and Rayyan Khan
Business Development Executive: Thilakh Rajavel

First published: March 2023

Production reference: 2050325

Published by Packt Publishing Ltd.
Livery Place
35 Livery Street
Birmingham
B3 2PB, UK.

ISBN 978-1-80461-322-1

www.packtpub.com

To my daughter, wife, and the Salesforce Trailblazer Community.

– Kristian Margaryan Jørgensen

Foreword

I have known and worked with Kristian Margaryan Jørgensen for over 4 years. Kristian is one of the most brilliant Salesforce experts that I have met. In this book, *Salesforce End-to-End Implementation Handbook*, Kristian takes you on a learning journey that introduces you to a framework for approaching Salesforce implementations end to end. The book walks you through the different Salesforce project phases – pre-development, development, rollout, and, finally, the continuous improvement phase.

In his presentations and examples, Kristian explains the required setup, activities, and organizational structures, pointing out potential difficulties you may encounter at each stage and sharing tips and tricks to avoid them. There are some included checklists as well that I found very handy. Kristian uses an example-driven method to explain his thoughts, which is easier to understand and relate to the reader.

With *this book*, you will learn the best practices for rolling out a successful Salesforce implementation. Once you have read Kristian's book, you will understand the different activities involved in each phase of the Salesforce project life cycle better. You will find this helpful regardless of the state of the Salesforce project(s) you are currently dealing with. The book also helps you spot and identify the gaps in your project's current setup that are causing delivery or post-delivery challenges.

In this book, Kristian encapsulates the knowledge gained through years as a world-class Salesforce consultant, a Salesforce architect, a community leader, and a passionate advocate. With his words, step-by-step instructions, diagrams, examples, and links to additional sources of information, you will learn how to enhance your skills and successfully and continually deliver Salesforce projects.

Tameem Bahri

Global engineering director of CRM at A.P. Moller - Maersk, Salesforce CTA, and the author of *Becoming a Salesforce Certified Technical Architect*

Contributors

About the author

Kristian Margaryan Jørgensen is a Salesforce solution architect at Waeg, an IBM Company. Kristian has nearly a decade of combined Salesforce end user, consultant, and solution architect experience. His experience from both the customer and consulting side of implementations enables Kristian to empathize when advising and challenging enterprise organizations how to plan, orchestrate, and scale their Salesforce implementations with a clear focus on usability and adoption to succeed in unlocking value from their Salesforce investments.

Kristian holds 14 Salesforce certifications, including Strategy Designer, Development Lifecycle and Deployment Architect as well as Application Architect, and System Architect. Kristian is also a certified SAFe Agilist.

Kristian lives in Copenhagen, Denmark, with his wife and daughter. In his spare time, he volunteers as group leader of the Salesforce Architect Group in Copenhagen, Denmark.

I want to thank my daughter, Esther, and wife, Sona, for their love and support.

I owe a tremendous thanks to Amanda Beard-Neilson and Antone Kom, as well as the dream team at Packt, for their efforts and patience in working tirelessly to improve the book.

And finally, I send deep gratitude to Tameem Bahri, who encouraged me when I shared the idea for this book.

About the reviewers

Antone Kom is a certified Salesforce application and system architect with over 15 years of experience building enterprise CRM solutions, 15 Salesforce certifications, and a deep understanding of Salesforce clouds and technologies. He helps organizations unlock the full potential of the Salesforce platform by delivering innovative and scalable solutions.

Amanda Beard-Neilson is a Salesforce MVP, a leader of the Salesforce London Admin Community Group, and a co-organizer of London's Calling Community Conference. She has numerous Salesforce certifications, is a multiple-star Trailhead Ranger, author of the Udemy course "Delivering Successful Salesforce Projects," and owner of the blog: `http://saasyabn.blogspot.com/`. For many years, Amanda has been a speaker at Dreamforce, Salesforce World Tours, and different Community Conferences. Find Amanda on Twitter: `@Amandbn1`.

Amanda first implemented Salesforce as an end user in 2007 and progressed her experience of the platform as a project manager for two Salesforce Platinum Partners. She returned to the end user environment in senior leadership roles within media, EdTech, and Fintech companies, specializing in digital transformation and change management.

Amanda now works for Capgemini, engaging in the successful delivery of Salesforce digital strategy for her customers.

Table of Contents

3

Determining How to Deliver Your Salesforce Project   47

4

Securing Funding and Engaging with Salesforce and Implementation Partners   69

5

Common Issues to Avoid in the Pre-Development Phase   93

Part 2: The Development Phase

6

Detailing the Scope and Design of Your Initial Release   105

10

Communicating, Training, and Supporting to Drive Adoption   167

11

Common Issues to Avoid in the Roll-Out Phase   183

Part 4: The Continuous Improvement Phase

12

13

14

Common Issues to Avoid in the Continuous Improvement Phase   241

Preface

Salesforce has been ranked the #1 CRM provider for the last 9 consecutive years. In their fiscal quarter ending July 2022, Salesforce overtook SAP as the world's largest provider of enterprise applications. Since its inception, Salesforce has expanded its initial sales domain-focused CRM offering by building and buying additional, complementary products – now spanning sales, e-commerce, service, field service, marketing automation, integration, RPA, analytics, AI, and much more.

With this huge growth and expansion, implementations of Salesforce have also seen a boom in business and technical complexity as well as organizational interconnectedness and dependencies, which all impact the way Salesforce projects and programs can be (and are being) delivered.

Literature is vast on Salesforce architecture and the technical aspects of developing Salesforce solutions (both with and without code), and there is no shortage of great implementation handbooks that cover *one* specific Salesforce cloud. Surprisingly, there is no book covering implementation lifecycle of a Salesforce project or program.

If your organization is going to invest – or already has invested – in Salesforce, you want to be sure to maximize the value from that investment. Value is unlocked when Salesforce is finetuned and adopted by your organization and is continuously delivering incremental value for your users and organization.

This book proposes a framework for how to approach Salesforce implementation projects and how to set your organization up to successfully manage a Salesforce program. Along the way, this book will provide you with insights into common issues and strategies to mitigate or prevent them altogether.

To be able to cover an arguably broad scope, this book is cross-functional and multi-dimensional. This means it covers many disciplines and domains without being a pure-play book on any one topic in particular.

The topics covered – all in a Salesforce context – include the following:

- Business architecture and strategy
- Enterprise architecture
- Salesforce architecture
- Project management
- Agile
- DevOps
- Governance

- Change management

- Program management

- Data management and analytics

- Product management

My hope is that this approach allows you to gain a holistic view of the nature of Salesforce projects and programs – without getting lost in the details.

By the end of this book, you will have gained the knowledge to be able to set Salesforce projects and programs up for success and deliver maximum value from investments.

Who this book is for

This book is for Salesforce consultants, architects, project and program managers, and product owners planning to implement Salesforce or already implementing Salesforce for their organizations. Salesforce consultants and delivery leaders at Salesforce implementation partners will also benefit from reading this book.

If you are pivoting your career to the Salesforce ecosystem – a vibrant, inclusive, extremely rewarding, and ever-evolving community – this book will give you a good understanding of what to expect when working on Salesforce projects and programs.

What this book covers

Chapter 1, *Creating a Vision for Your Salesforce Project*, begins with an introduction to the lifecycle of Salesforce implementations and gives you an overview of the pre-development phase of a Salesforce project. It then dives into strategy analysis and methods to help you understand your company's imperative for change. This leads you to be able to define the vision for your salesforce project – your *why*.

Chapter 2, *Defining the Nature of Your Salesforce Project*, takes you through the activities you will be carrying out when defining the scope of your Salesforce project. We will also look at how you can – at this early stage – begin to consider how that translates to a high-level technical scope for your Salesforce solution and, critically, what technical enablers should typically be considered in it. This will allow you to define the *what*.

Chapter 3, *Determining How to Deliver Your Salesforce Project*, explores the topics of delivery methodology, change management, and governance. You will learn about what aspects to consider when deciding the right delivery methodology for your project, and how to distinguish between projects and programs. We'll also discuss the advantages and disadvantages of various approaches for project phasing. We'll cover what changes typically result from a Salesforce project, and what you should do to manage it. Lastly, we'll guide you in laying the foundation for your overall Salesforce governance – your Salesforce Center of Excellence. The chapter enables you to answer the *how* of your Salesforce project.

Chapter 4, Securing Funding and Engaging with Salesforce and Implementation Partners, covers the financial aspect of a Salesforce project, along with engaging with external parties. You'll learn how to create a business case based on business objectives and KPIs aligned with the vision for your Salesforce project. We'll cover both the revenue-driving and efficiency/cost-saving impacts of your Salesforce project and program. You'll be able to create and present your business case including the payback time and return on investment for your Salesforce project. Next, we'll cover how you can engage with Salesforce to request a quote and be further inspired. Lastly, we'll cover engaging and contracting with an implementation partner, and what to look for when choosing one. This chapter allows you to answer *how much* your Salesforce project or program will cost – and how much your organization stand to gain from it.

Chapter 5, Common Issues to Avoid in the Pre-Development Phase, introduces issues and root cause analysis. We will deep-dive into common issues faced in the pre-development phase of a Salesforce project and offer strategies for mitigation and prevention. Finally, we will go through a summarized checklist for the key activities you should have covered in the pre-development phase – enabling you to evaluate the state of your Salesforce project.

Chapter 6, Detailing the Scope and Design of Your Initial Release, begins by providing an overview of the development phase of your Salesforce project. In the chapter, we cover the activities and deliverables associated with detailing the scope of your Salesforce project. We cover the key architecture artifacts you should understand and have created for your Salesforce project. We also discuss levels of solution design and how much effort to put into creating it upfront before commencing development. We also cover aspects to consider when evaluating user story solution designs.

Chapter 7, Building and Testing Your Initial Release, covers the intricacies of getting the development of your Salesforce project off to a great start –through to completion. We explore the roles and responsibilities of your team members, how to plan your build phase, and considerations for setting up your Salesforce development model and testing environments. We go through the development process and highlight the mechanisms you should implement to govern your Salesforce project in the development phase. We also discuss the steps to take when preparing your data migration plan.

Chapter 8, Common Issues to Avoid in the Development Phase, presents common issues in the development phase of a Salesforce project and offers strategies for mitigation and prevention. Finally, we go through a summarized development phase checklist, enabling you to evaluate the state of your Salesforce project.

Chapter 9, Deploying Your Release and Migrating Data to Production, gives you an overview of the roll-out phase of a Salesforce project. We go through the practicalities of deploying your Salesforce solution through environments and how to drive user acceptance testing. Finally, we cover how to determine your preparedness ahead of data migration and signing off to go live.

Chapter 10, Communicating, Training, and Supporting to Drive Adoption, covers the critical activities for planning, communicating, and managing change. We'll discuss and provide templates for phasing your rollout, as well as different ways to deliver training. We'll look at how you should provide hyper care and when you can safely transition to ongoing production support. We'll explore ways in which you should be monitoring and driving the adoption of your Salesforce solution to maximize the value derived from your investment.

Chapter 11, Common Issues to Avoid in the Roll-Out Phase, presents common issues in the roll-out phase of a Salesforce project and offers strategies for mitigation and prevention. We also go through a summarized checklist for the roll-out phase, enabling you to evaluate the state of your Salesforce project.

Chapter 12, Evolving Your Salesforce Org and DevOps Capabilities, begins by providing an overview of the continuous improvement phase of your Salesforce program. We explore the concept of a product organization and why you should aspire to transform your organization to become one. We'll describe how you can implement suitable governance for your Salesforce program and platform through an updated Salesforce CoE and core roles in it. Lastly, we'll dive into Salesforce DevOps and explore ways in which you can evolve your Salesforce DevOps capabilities to deliver greater value for your users and organization.

Chapter 13, Managing Your Salesforce Data to Harvest the Fruits of Customer 360, covers the important domain of data management and governance. We'll go through fundamental concepts such as data categories, the data lifecycle, and dimensions of data quality. We'll show you how you can govern, monitor, and improve the quality of your Salesforce data. We then guide you through an assessment of your own analytics initiatives and describe how you can consider leveraging your Customer 360 data to maximize the return on your Salesforce investments.

Chapter 14, Common Issues to Avoid in the Continuous Improvement Phase, presents common issues typically faced in the continuous improvement phase of a Salesforce program and offers strategies for mitigation and prevention. We also go through a checklist of what you should do when initially entering the continuous improvement phase, as well as what you should continually be doing to further unlock value for your users and organization.

To get the most out of this book

If you have some familiarity with basic business, CRM, and project management concepts, it will help you to get the most out of this book. Professional working experience with Salesforce is an advantage, but not a prerequisite to reading this book.

While Salesforce development and technical architecture are more well-defined domains, Salesforce delivery is less so. That is what this book aims to do – to offer a framework and a perspective.

I have been part of driving Salesforce implementations primarily for European companies operating locally and globally. Your views and experiences likely differ from mine. I hope to start a conversation and invite you to lean in and offer your perspectives on *#BetterSalesforceDelivery*.

Let's go!

Download the color images

We also provide a PDF file that has color images of the screenshots and diagrams used in this book. You can download it here: `https://packt.link/CIbLO`.

Conventions used

There are a number of text conventions used throughout this book.

Bold: Indicates a new term or an important word. Here is an example: "Your ability to evolve your **Salesforce org** to changing business needs, as well as managing and leveraging your data, is your top activity in this phase."

> **Tips or important notes**
> Appear like this.

Get in touch

Feedback from our readers is always welcome.

General feedback: If you have questions about any aspect of this book, email us at `customercare@ packtpub.com` and mention the book title in the subject of your message.

Errata: Although we have taken every care to ensure the accuracy of our content, mistakes do happen. If you have found a mistake in this book, we would be grateful if you would report this to us. Please visit `www.packtpub.com/support/errata` and fill in the form.

Piracy: If you come across any illegal copies of our works in any form on the internet, we would be grateful if you would provide us with the location address or website name. Please contact us at `copyright@packt.com` with a link to the material.

If you are interested in becoming an author: If there is a topic that you have expertise in and you are interested in either writing or contributing to a book, please visit `authors.packtpub.com`.

Share Your Thoughts

Once you've read *Salesforce End-to-End Implementation Handbook*, we'd love to hear your thoughts! Scan the QR code below to go straight to the Amazon review page for this book and share your feedback.

https://packt.link/r/1-804-61322-3

Your review is important to us and the tech community and will help us make sure we're delivering excellent quality content.

Download a free PDF copy of this book

Thanks for purchasing this book!

Do you like to read on the go but are unable to carry your print books everywhere?

Is your eBook purchase not compatible with the device of your choice?

Don't worry, now with every Packt book you get a DRM-free PDF version of that book at no cost.

Read anywhere, any place, on any device. Search, copy, and paste code from your favorite technical books directly into your application.

The perks don't stop there, you can get exclusive access to discounts, newsletters, and great free content in your inbox daily.

Follow these simple steps to get the benefits:

1. Scan the QR code or visit the link below

https://packt.link/free-ebook/978-1-80461-322-1

2. Submit your proof of purchase
3. That's it! We'll send your free PDF and other benefits to your email directly

Part 1: The Pre-Development Phase

This part will guide you through the activities before commencing development. We will also introduce **Packt Manufacturing Equipment** (PME), the scenario company we'll be following throughout the book.

This part has the following chapters:

- *Chapter 1, Creating a Vision for Your Salesforce Project*
- *Chapter 2, Defining the Nature of Your Salesforce Project*
- *Chapter 3, Determining How to Deliver Your Salesforce Project*
- *Chapter 4, Securing Funding and Engaging with Salesforce and Implementation Partners*
- *Chapter 5, Common Issues to Avoid in the Pre-Development Phase*

1

Creating a Vision for Your Salesforce Project

To start a journey, you need to determine where you want to go. This chapter guides you through methods of analysis to understand your company's imperative for change, which will be the foundation for you to be able to create a vision for your **Salesforce project**.

After creating a vision for your Salesforce project, you'll then be ready to move on to the next activities in the pre-development phase of your Salesforce implementation.

In this chapter, we'll cover the following main topics:

- Introducing the life cycle of Salesforce implementations
- An overview of the pre-development phase of your Salesforce project
- Understanding your company's imperative for change
- Defining the vision for your Salesforce project
- Iterating in the pre-development phase

Let's go!

Introducing the life cycle of Salesforce implementations

This book will provide a framework for approaching the entire implementation life cycle specific to Salesforce projects – from **project inception** in the pre-development phase to the continuous improvement phase. Here's a brief overview of the phases involved:

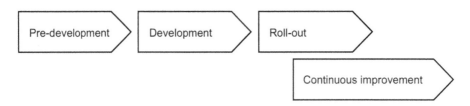

Figure 1.1 – The project phases of a Salesforce implementation

In *Figure 1.1*, you can see each phase of a Salesforce project. Let's briefly define each phase:

- **Pre-development phase**: In this phase, you are creating a vision for your Salesforce project, defining its nature, determining how to deliver it, securing a budget for it, and engaging with Salesforce and implementation partners. In this phase, you will also be setting up the governance body, which will own and oversee the delivery of your Salesforce project.

- **Development phase**: This phase begins with a project kick-off event, following which the project team will start delivering the project. There are different methodologies for delivering Salesforce projects, which we will cover later in *Chapter 3, Determining How to Deliver Your Salesforce Project*. Whichever methodology you choose, the development phase is where your Salesforce solution's scope is detailed, designed, built, and tested to ensure quality.

- **Roll-out phase**: In this phase, you are deploying your release, migrating data, training your users, providing support, and closely monitoring how users are adopting your solution and responding accordingly.

- **Continuous improvement phase**: Finally, you will find yourself in a state of continuous improvement. Your ability to evolve your **Salesforce org** to changing business needs, as well as managing and leveraging your data, is your top activity in this phase.

Salesforce projects – especially **greenfield implementations,** where an organization sets up their **Salesforce org** for first-time use – do not start at the project kick-off event. Neither should Salesforce projects end directly after user training in the roll-out phase.

Now we have understood the overall phases of a Salesforce project. Let's dive into the pre-development phase of your Salesforce project.

An overview of the pre-development phase of your Salesforce project

Each phase of a Salesforce project has its own set of **key milestones**, along with activities to reach the milestones.

Let's get started with key milestones of the first phase of your Salesforce project.

Key milestones of the pre-development phase

The pre-development phase of your Salesforce project starts at inception. Inception is defined as the point where your company has made a choice of technology from a range of CRM providers, and Salesforce has been chosen as your future **CRM platform**. Deciding on a CRM provider is outside the scope of this book.

The key milestones of the pre-development phase are securing an initial budget for your Salesforce implementation and contracting with **Salesforce Inc.**, and most often an **implementation partner**, to assist you in the development and roll-out of your solution.

To reach these milestones, several activities are required. We will go over these activities in the next section.

Activities in the pre-development phase

The activities in the pre-development phase of your Salesforce project are illustrated in the following figure:

Figure 1.2 – Activities in the pre-development phase

There are two key points to note from *Figure 1.2*:

- Each activity in the pre-development phase is covered by a chapter in this book, so you can easily navigate to the part you want to dive into.

- There are no arrows in the diagram. While the order of the activities generally means you should go clockwise through the process, it is not always the case. We will come back to this point in the *Iterating in the pre-development phase* section of this chapter.

Now that we have established *what* the overall activities consist of in the pre-development phase, let's look at *who* will be performing the activities.

Introducing the Salesforce taskforce

To carry out the activities in the pre-development phase, the executive sponsor of the Salesforce implementation should create a **Salesforce taskforce**.

The Salesforce taskforce should consist of three key roles, with representatives from the business organization, the IT organization, and your **Project Management Office (PMO)**. Finally, the Salesforce taskforce *must* have an executive sponsor.

The roles and responsibilities of each member are described in *Table 1.1*:

Taskforce member role	Taskforce member responsibilities
Business owner	• Lead the analysis of the company's strategic situation • Determine required business capabilities • Determine business KPIs to measure the success of the project
Tech owner	• Based on the required business capabilities, determine high-level target technical scope • Ensure target technical scope aligns with enterprise architecture principles and plans
PMO representative	• Taskforce project management • Stakeholder management • Set meetings with internal and external stakeholders, capturing minutes and actions • Set and schedule taskforce meeting cadence, and report to the executive sponsor • Escalate issues and decisions not able to be taken by the taskforce
Executive sponsor	• Support and coach taskforce • Act as an escalation point

Table 1.1 – The responsibilities of taskforce member roles

Duration of the pre-development phase

During the beginning of the Covid-19 pandemic, some companies realized their **imperative to change** was urgent and turned around all the activities in the pre-development phase in a matter of a few weeks. While it may have rescued small- and medium-sized businesses in exposed industries, you should not rush through this phase.

The solid preparation and organizational alignment work done in the pre-development phase will make your Salesforce project more robust and have a greater chance for success, compared to a rushed pre-development phase.

Let's continue to the next section where we will uncover your company's need for change.

Understanding your company's imperative for change

In this section, we will analyze your company's situation through various methods to be able to determine the vision for your Salesforce project.

To help exemplify the methods used, let us first introduce the scenario company that we'll follow throughout this book.

Introducing our scenario company – Packt Manufacturing Equipment (PME)

Let's look at some initial facts about the scenario company:

- Company name: Packt Manufacturing Equipment (PME)

- Industry: Manufacturing

- Annual revenue: 300 million USD

- Number of employees: 2,000 FTEs

- Geography: Operating and selling globally; direct in mature PME markets, and through distributors in emerging PME markets

Meet the PME Salesforce taskforce charged with shaping the project in the pre-development phase:

- **The business owner**: *Kalinda Keen. Kalinda is a manufacturing industry veteran. She has held multiple roles, in many functions, within PME. Currently, she is working in the global commercial excellence department, responsible for building capabilities for local marketing, sales, and customer service organizations. Kalinda is closely aware of the market trends and forces affecting PME.*

- **The tech owner**: *Cary Sharp. Cary is a rising star in the global IT department. Cary is a solution architect transitioning to an enterprise architect position and is extremely interested in cloud computing. He has worked at PME for 4 years and is well aware of the PME system landscape.*

- **The PMO representative**: *Alicia Fast. Alicia is new to PME, having joined recently from another industry. She has lots of project management experience, and her recent job involved managing a Salesforce implementation.*

- **The executive sponsor**: *Diane General. Diane is a* **Senior Vice President** **(SVP)** *of global commercial excellence, reporting to the CEO of PME. Kalinda reports to Diane. Diane has been with PME for over 10 years, in multiple managerial positions ranging from country leadership to functional management roles. She is well liked and trusted within the organization and has been part of PME's board committee on digital transformation strategies.*

After getting to know the scenario company, PME, let's return to your company.

Your company's environment, situation, and challenges

To lay out a path, in addition to knowing where you want to go, you also need to understand where you are coming from and, critically, why you need to move. Your first activity in the pre-development phase is, therefore, to analyze and establish your **company's situation**.

In the following section, we will cover basic methods of understanding how your company is organized, who your customers are, and the products and services you sell.

> **Important note**
> Your role as part of the Salesforce taskforce should not be to produce the analysis output, but *to consolidate and understand what over arching forces drive your company's imperative for change.*

Understanding how your company is organized

Before you endeavor to perform any strategy analysis, you should first seek to establish how your company is organized. The simplest way to achieve this is by looking at your company's org charts. *Charts* is a plural intentionally, as there will be many layers.

You will want to understand the following regarding your company's structure:

- Your local organizations' structure:

  - How many countries does your company directly operate in?

  - How are the countries or local organizations typically organized?

    - Does each country have its own leadership team or does each country's sales, customer service, and marketing team report directly to group sales, customer service, and marketing management, respectively?

    - How much variation is there among the structures of your local organizations?

- How many people are employed in each country or sales organization?

  - How many are employed per country or sales organization?

  - How many are employed in marketing, sales, and customer service functions, respectively?

  - Beyond sales, marketing, and customer service, which other functions do local country or sales organizations manage, and to what extent? This could be back-office sales support functions, local logistics, and local IT support.

- Your group-level organization:

  - This would typically be an org chart of your company's group management

  - Pay special attention to IT governance:

    - Is group IT responsible for setting guidelines and providing ad hoc advice, or do they firmly own the enterprise IT landscape and work to ensure compliance? It may be somewhere in the middle but be sure to get a good understanding.

- Your partner, reseller, or distributor network:

  - If your company sells its product or services through partners, how are the relationships with those partners managed?

  - Are the partner relationships managed at the group level or by the local sales organization? Or at some intermediary level?

  - How many partners does your company have agreements with?

  - What countries do your partners cover?

  - Do your partners sell *all* your company's products and/or services or only a subset?

Tip

Engage with your HR department to understand how your internal company is structured. Regarding your partner network, you may have to do some investigation internally to find out who manages those relationships.

Be aware that organizations are dynamic and that small organizational changes happen frequently, and larger changes also occur from time to time.

To understand how an org chart could look, let's look at PME's company structure in the following diagram:

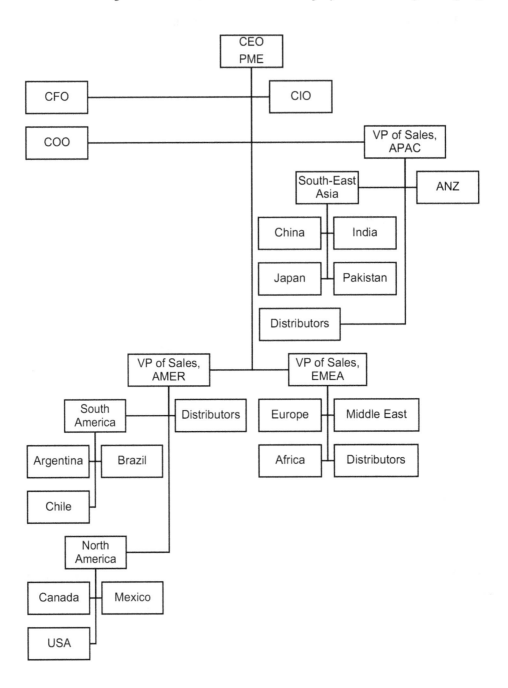

Figure 1.3 – PME org chart

In Figure 1.3, you see the CEO of PME at the top. Reporting directly to the CEO are the CFO, COO, and CIO. At PME, also reporting directly to the CEO are the SVPs of each of the three PME regions: AMER, APAC, and EMEA. Reporting to each regional SVP are the local country organizations of PME; some have a country leader responsible for multiple smaller countries. The regional SVPs at PME are also responsible for managing the relationships with the distributors operating in the regions.

You have now established how your company and partners are organized. Let's move on to understand to whom your organization sells, and what you sell.

Understanding who your customers are

You should seek to understand the following key aspects of your *customers*:

- How many customers do you have?

- Do you have a few large customers that make up the bulk of your revenue or is revenue spread across many customers? Is there a change in this split?

- How large are your customers typically in terms of revenue and number of employees? Seek to understand your median customer revenue.

- What industries do your customers belong to?

- Who are your customers' end customers or consumers?

Understanding the products and services you sell

You should seek to understand the following key aspects of your *products and services*:

- What challenges do your products help your customers overcome?

  - If you have many product categories and/or customer segments, you should attempt to consolidate the answers into three to four key groupings for clarity and comprehension

- How many products do you have?

  - Do you have a few products that make up the bulk of your revenue or is revenue spread across many products?

- Are your products and services finished, pre-packaged products and/or services, or are they highly configurable?

  - Ensure you also determine whether your company has any plans for making your products and services configurable in the future

> **Tip**
>
> Engage with sales and marketing leaders at both your company's local and group levels to get an understanding of your customers and how your products and services add value to them. You may also request help from your finance or BI departments for data and insights about your customers.

Introducing the basics of strategy analysis

While the literature on business strategy is vast and growing, some fundamental analysis of your company is needed to understand its situation and the imperative for change. In this section, we will cover the most used methods of analysis.

> **Tip**
>
> If your company has its own framework of strategy analysis, you should utilize insights from that strategy analysis to understand your company's situation.

PESTLE analysis

The decade of the 2020s has begun with economic turmoil following the Covid-19 pandemic, supply chain pressures, a new war in Europe, continuing wars in other parts of the world, and inflation. These are all macro-level phenomena that – in varying ways, depending on your industry and geography – impact your company's situation.

Using **PESTLE analysis**, you categorize these impacts into six factors:

- **Political**: Political factors include any potential or realized political decisions that are impacting or may impact your company, such as taxes, subsidies, and reforms. *Your executive sponsor and senior management can quickly list the key current political trends they are keeping an eye on.*

- **Economic**: These factors are numerous; some key examples are inflation, interest rates, consumer confidence, business confidence, and GDP growth of the markets your company serves. *Your CFO or finance business partner is your go-to source for input on economic factors influencing your company.*

- **Social**: Social factors cover the demographic makeup and trends of your markets' populations, as well as their cultural preferences. *Your marketing department should have great insights into shifting trends in your target demographic – consumer- or business-wise.*

- **Technological**: These factors include advances in technology that may enable you (and your competitors) to produce and deliver your products and services more efficiently at a lower production cost, as well as being able to market and sell your products and services more effectively. Common examples include automation in manufacturing, and more efficient truck engines or electric vehicles. Leveraging customer insights for better segmentation and targeting to market and selling more effectively are other examples of technological factors. *Talk with your company CTO or head of innovation to gather these insights.*

- **Legal**: Legal factors include legal precedents relevant to your industry and operating geographies, as well as any specific legal issues and risks your company may be exposed to. *The general counsel of your company, your CFO, or the head of risk management will be able to discuss the legal factors with you, though you will not be able to gain insights into all areas as some may be confidential.*

- **Environmental**: These environmental factors include key trends of the physical environment in the geographies your company operates in. Obvious examples are climate change and the direct and indirect effects of climate change. Although the direct impact of climate change may have less impact on your company, the indirect effects, such as political instability, change in consumer and business preference, and shifts in product and service demand may have a greater influence on your company's situation. *Your executive sponsor and senior management can tell you whether any environmental factors are particularly influential to your company.*

Having established the macro factors influencing your company, let's move on to understand the competitor situation of your company.

Porter's five forces

This is a standard method created by Michael E. Porter, the renowned Harvard Business School professor of strategy and competition, for analyzing the forces that impact a business's competitive situation.

Figure 1.4 – Porter's five forces

Let's go through each of the five forces:

- **Industry rivalry**: You should seek to establish answers to these key questions when describing your company's industry rivalry:

  - **Nature of your industry's rivalry**: How are your company and your competitors competing? Examples include price cuts, guerilla marketing methods, and locking in customers with exclusivity contracts.

  - **Competitive situation**: Is your company a dominant player or one of many competitors in your operating markets? Who are the biggest competitors? Who is gaining and losing market share? Use the last reporting year versus the previous reporting year to see this trend.

- **Threat of new entrants**: Are there any indications of or actual new entrants entering your industry? What is their operating model and means of differentiation?

- **Threat of new substitutes**: Are customers using alternative products or substitute products in new ways to meet their needs?

- **Bargaining powers of suppliers**: If your company (and general industry) is relying increasingly on a smaller number of suppliers, then they will have an increase in bargaining powers with the result being increased prices.

- **Bargaining powers of buyers**: Similar to the bargaining powers of suppliers, if you sell to a highly concentrated group of customers, your company will have less bargaining power.

Market development

Determine whether the markets your company is operating in are growing, stagnant, or declining. Compare the **market developments** of your key markets. Also, note whether there are any cyclical patterns and seasonality in your market. *Your CFO or finance business partner is your source of information here.*

Company's financial performance

You may easily have access to these key figures from your company's latest **annual report** (for example, if your company is publicly traded). If not, work with your executive sponsor and their finance business partner to get a perspective on your company's financial situation.

- **Top-line revenue growth**: This tells you whether your company is growing or not. If it is growing, great! But seek to understand this in the context of the overall market growth. If your company is growing slower than the market, your company is losing market shares to the competition. You should then seek to understand the reason for this. *The leaders of your sales and marketing organizations should be able to help you determine this.*

- *Your finance business partner will be able to help you understand the drivers for revenue growth.* Ask for a **price-volume-mix (PVM) analysis** for the revenue growth of last year versus the previous year to understand whether your revenue is primarily growing due to more products or services being sold, higher prices, or a change in the mix of the products and services sold.

- **Gross margin %**: Determine whether your gross margin percentage is increasing or decreasing. This is a sign of your *company's viability*. Compare your company's gross margin % with your key competitors. If it is lower than your competitors, your competitors are either able to command higher prices for their products and services or are more efficient in producing their products at a lower cost, or both.

  - Your finance business partner will be able to help you understand the drivers for change in gross margin % using the same PVM analysis method

- **Operating profit %**: Determine whether your company's operating profit % is increasing or decreasing. This figure is the result of multiple factors in your company's profit and loss statement. It is a sign of your *company's financial sustainability*.

  - Request the help of your finance business partner to interpret – and put into perspective – your company's operating profit %

Analysis of your company's financial performance is the last component of your strategy analysis. Next, we will put all the analysis together.

Summarizing your company's strategic situation

The last step of your strategy analysis is to consolidate and structure the key, most important, and influential facts, factors, and trends into an easily comprehensible overview. Let's explore the classic **Strengths, Weaknesses, Opportunities, and Threats (SWOT)** analysis model:

- **Strengths**: What are the key strengths of your company that make you stand out from the competition?

  - *Review your findings in Porter's five forces analysis as well as your company's financial performance analysis for input on this*

- **Weaknesses**: What are the key weaknesses of your company that make you vulnerable to the competition?

  - *Review your findings in Porter's five forces analysis as well as your company's financial performance analysis for input on this*

- **Opportunities**: What are the key opportunities present for your company to pursue?

  - *Review your PESTLE analysis as well as your market development analysis for input on this*

- **Threats**: What are the key threats against your company that your company should actively mitigate?

  - *Review your PESTLE analysis as well as your market development analysis for input on this*

To understand what a SWOT could contain, let's have look at the SWOT analysis of our scenario company, PME:

- **Strengths**:

  - *Trusted brand by existing customers*

  - *Large and experienced manufacturing capabilities to produce superior products versus competition*

- **Weaknesses**:

  - *Company's financial performance. Top-level revenue growth is slower than the competition – losing market share. PVM's driver is simply volume*

  - *Lack of modern system support for a modern commercial organization*

  - *Declining customer experience and* **customer satisfaction** *(***CSAT***) compared to the competition*

- **Opportunities**:

  - *Turning the service department into a profit center by commercializing service offerings*

  - *Several sub-segments in various geographies that are currently unserved*

- **Threats**:

  - *Existing and new competitors are rapidly developing new ways to modernize their companies' go-to-market strategies, thus gaining more market shares. The competition is providing customers with automated and personalized marketing communication, seamless omnichannel buying experiences, digital self-service, and predictive maintenance solutions.*

  - *Some competitors are also leveraging the* **Internet of Things** *(***IoT***) and* **artificial intelligence** *(***AI***) technology to provide advanced predictive and prescriptive insights to enhance their customer's experience.*

Having gathered insights into your company's strategic position, now it's time to put the insights to use to create a vision for your Salesforce project.

Defining the vision for your Salesforce project

By consolidating your strategy analysis, you should have a good understanding of the potential purposes of your Salesforce project.

To make it crystal clear to yourself, the Salesforce taskforce, and eventually your wider organization, make sure you can explain why *now* is the right time for your Salesforce project.

Why now?

What is the main reason(s) why the Salesforce project is being considered *now*?

Common reasons for implementing the Salesforce CRM include the following:

- Loss of competitive advantage:

  - Competitors are gaining market shares by providing a superior customer experience in the buying process and your current customers are demanding to be better served

    - Your market analysis, competitor analysis, and SWOT should indicate whether this is the main reason

- **Legacy CRM** system reasons:

  - A home-grown CRM system is not maintainable/upgradable to support your company's digital transformation ambitions

    - Your analysis (or your enterprise architect's analysis) should have uncovered this

  - Legacy CRM system licenses and support expiring: this could either be due to a decision by your company to not continue with the existing CRM system, or it could mean your current provider decided your (on-premises) CRM system is reaching the end of its life

- No system support:

  - Your company has no CRM system to support your business processes: your business users are using email and spreadsheets to manage their marketing, sales, and customer service operations

For our scenario company, PME, the main reason why now is the right time for the Salesforce project is that the current CRM system is not adequate for the digital transformation ambitions of PME's leadership. In addition to that, several smaller countries are not utilizing the CRM system provided by PME as they find it cumbersome and unsatisfactory, altogether. Instead, they are resorting to emails and spreadsheets, and some have even acquired licenses for other smaller CRM providers and email marketing providers (much to the dismay of the IT team). Finally, customer churn is increasing, CSAT scores are decreasing, and customer service is overwhelmed with customer calls and emails.

Your company may well have other valid reasons for why *now* is the right time to embark on a new CRM journey. Whatever the reason is, make sure you fully understand it and that your wider stakeholder group is aligned with them.

Having concluded why *now* is the right time for your Salesforce project, let's move on to craft the vision for your Salesforce project.

Creating a vision for your Salesforce project

To make sure your **vision statement** will live up to its purpose, let's first go through the characteristics of a Salesforce project vision statement.

A vision statement should fit the following criteria:

- Is a short, single-sentence paragraph
- Describes the essence and goal of your Salesforce project
- Should not be technical
- Should be easy to understand for your stakeholders
- Is addressing your employees as the audience
- Should create excitement about the future and the Salesforce project

You and your Salesforce taskforce may make several draft vision statements before settling on the right one.

> Tip
> Have sessions with your executive sponsor to ensure you get this right.

The Salesforce taskforce at PME has gone through several rounds of information gathering to create a consolidated view of PME's situation, challenges, and reasons for why now is the right time for their Salesforce project. They have also come to the end of the first main activity in the pre-development phase: they have created a vision for their Salesforce project aligned with their stakeholders:

> PME's **Salesforce CRM** *project will* **enable us to focus** *on value-adding,*
> *customer-facing activities across* **marketing, sales, and customer service**, *which*
> *are* **stimulating to us** *while* **delivering delightful experiences for**
> **our customers.**

Let's break down the anatomy of PME's vision statement. Firstly, it makes it clear *who* the project is intended for, namely employees across marketing, sales, and customer service. It also explicitly states *what* the project will enable PME to do, that is, to focus on value-adding, customer-facing activities, as opposed to the current state of a cumbersome, manual CRM system. Finally, PME's vision statement addresses *why* the project is happening: to stimulate PME employees and delight customers.

With a clear vision for your Salesforce project created, you are ready to move on to the next chapter of your Salesforce journey.

Iterating in the pre-development phase

So far, you have been engaging with internal stakeholders of your company to understand it and to create a vision for your Salesforce project. Internal stakeholders are essential when defining a Salesforce project. As mentioned earlier, as you progress through the activities of the pre-development phase, you will likely come back and revise your organizational overview and strategy analysis as you become more knowledgeable and uncover new insights about your company and its environment. You may also find yourself tweaking the vision statement of your Salesforce project, and that's okay as long as it is not a 180-degree change.

Therefore, be mindful not to set things in stone as ultimate truths and holy decisions. Keep an open mind as you progress through the first iteration of the pre-development phase.

Summary

You have come a long way since starting this chapter. You first learned about the phases in the implementation life cycle of Salesforce projects. Next, you were introduced to the key milestones, outcomes, and activities in the pre-development phase of your Salesforce project. Then, you made the effort to understand and describe your company's situation and imperative for change, enabling you to create a vision for your Salesforce project.

Having established a vision for your Salesforce project – the *why* – you are ready to move on to the next chapter of your Salesforce implementation, *Chapter 2, Defining the Nature of Your Salesforce Project.*

2
Defining the Nature of Your Salesforce Project

In this chapter, you will learn how to scope the desired business capabilities of your Salesforce project and determine the high-level hypothetical technical scope required to support those capabilities. These two elements together will be referred to as the nature of your Salesforce project.

You will be introduced to business process mapping and designing workshops for understanding current business processes and defining the business capabilities of your Salesforce solution. You will then be introduced to the Salesforce license types and learn how to determine the high-level hypothetical technical scope for your Salesforce solution.

We will continue to follow our scenario company, **Packt Manufacturing Equipment** (**PME**), and progress through the activities to define the nature of their Salesforce project.

After the nature of your project is defined, you'll be ready to move on to the next activities in the pre-development phase of your Salesforce implementation.

This chapter will cover the following main topics:

- Defining the capabilities to be supported by your Salesforce solution
- Determining the high-level technical scope of your Salesforce solution
- Iterating in the pre-development phase

Let's go!

Defining the capabilities to be supported by your Salesforce solution

In the previous chapter, you established your company's situation and the overall purpose and vision for your Salesforce project. As illustrated in the following diagram, it is now time to get more specific and define the nature of your Salesforce project:

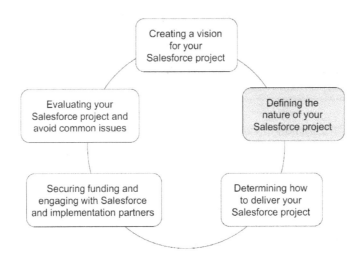

Figure 2.1 – Activities in the pre-development phase

Before jumping into defining the business capabilities for your Salesforce project, let's first understand what business capabilities are and why they are important to define.

Understanding business capabilities

All companies and organizations have a number of **business capabilities**. But what are they really? A business capability is comprised of **people** with a skillset or area of competency who carry out work according to a **business process**, supported by one or more **systems** while having access to and the ability to create and update relevant **data** in order to achieve desired **business outcomes**.

An example of a common business capability is **account and contact management**. Account and contact management covers a company's ability and processes for managing customer and customer contacts' data.

As you may have figured out, business capabilities are important to understand and map for a company *before* being able to determine what the technical scope should be. The reason is that systems are a component of a business capability. Knowing what business capabilities and business outcomes your Salesforce solution should support is *critical* before diving into technical architecture and design.

A business capability may be achieving the desired outcomes to a smaller or greater extent. We say business capabilities *vary in maturity* so let's explore business capability maturity.

Understanding business capability maturity

To understand the concept of maturity of business capabilities, let's look at a simple example.

If your company provides any kind of support for your customers, you implicitly have a *customer service* domain business capability. The exact business capability could be customer service ticket management or **case management**, as it is often referred to in a Salesforce solution. You may currently have little or no system support beyond a customer service phone number and people picking up the phone to answer customer inquiries. The customer service staff may have received little or no structured training in a predefined process and have no access to any data to help them provide customer service. In this case, your current case management business capability could be categorized as having *low maturity*.

On the other hand, your customer service capability may consist of well-trained and skilled people who work according to a predefined process. They may also be supported by an aligned and fit-for-purpose system that provides adequate data insights to help your customer service staff in providing top-notch customer service. If these descriptions are valid, your case management business capability would be considered to have *higher maturity*.

Next, let's look at categorizing business capabilities.

Categorizing business capabilities

Business capabilities are grouped by **domain**, where a domain represents an overall function of a company, such as marketing, sales, customer service, operations, finance, IT, and HR.

Determining up front what overall business domains are in the scope of your Salesforce project is critical before diving into the specifics of business capabilities.

In addition to grouping your company's business capabilities by domain and assessing the maturity of each, you should be aware that every company is different. Different domains and business capabilities exist from industry to industry and within a company's different business lines and geographies and are therefore important to identify on a case-by-case basis. It is important to understand which business capabilities are considered strategic within your company and therefore given focus and resources to expand and mature – in order to achieve target business outcomes.

As you progress through the activities of determining your current business capabilities, keep in mind what you have learned about the maturity of business capabilities and seek to identify which capabilities play a strategic role in the success of your company.

Now let's take a look at the role of the Salesforce taskforce in defining business capabilities for your Salesforce project.

The role of the Salesforce taskforce in defining business capabilities for your Salesforce project

As a member of the Salesforce taskforce, you will be deeply engaged in carrying out activities to determine the business capabilities in the scope of your Salesforce project.

Both the business owner and tech owner will be busy creating and consolidating the output and artifacts produced during these activities.

Determining your high-level business processes

The first step is to create a high-level overview of your company's overall business processes. An appropriate illustration method is a simple flow chart of the end-to-end process from marketing activities through to invoicing and after-sales customer service. In *Figure 2.2*, you'll find an example of a simple flow chart of PME's high-level business processes:

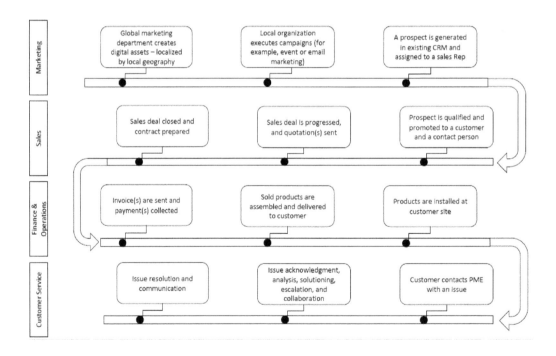

Figure 2.2 – PME's current high-level business process

This simple flow chart is suitable at this stage as it does not allow for branching or decisions, forcing you to stay at a high abstraction level. The high abstraction level also means most, if not all, stakeholders you show it to will be able to understand and relate to it.

Next, we will look at breaking the high-level business processes into business capabilities.

Creating an overview of your current business capabilities

To determine your current business capabilities, start by determining the overall business functions, the **domains**. Next, go through each step in the high-level business process chart you created and break each step into separate business capabilities. Keep the definition of business capabilities in mind as you name your business capabilities – distinguish between business processes where new people, systems, and data are introduced.

> **Tip!**
> Engage with an enterprise architect at your company to validate your overview of business capabilities. Make updates as required.

Once you have all your company's capabilities determined, create a business capability map for a simple illustration.

> **Tip!**
> Salesforce is *fundamentally and primarily* – among many other features – a CRM system. Therefore, you should focus on your business' marketing, sales, and customer service domain capabilities as a starting point for your Salesforce project.

In the following diagram, you can see PME's business capability map – focused on customer relationship management business capabilities:

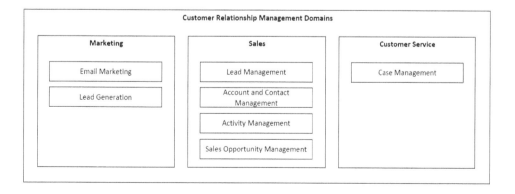

Figure 2.3 – PME's current business capability map

In *Figure 2.3*, PME's business capability map shows the business capabilities currently supported by PME's current CRM **tech stack**. Tech stack means the systems used by one or more marketing, sales, or customer service persons to execute their work.

Having determined the high-level business processes of your company, let's move on to better understanding the identified capabilities and get your end users' perspectives.

Facilitating workshops to understand current key business processes

Up until this point, you, the Salesforce taskforce, will have carried out desk research about your company's overall business processes and engaged with an enterprise architect and your executive sponsor. To validate your understanding of your company's CRM business capabilities, it's critical to get your end users' perspectives.

> **Tip!**
> Approach your Salesforce project from a change management perspective – early involvement of your end users in your Salesforce project will lay the groundwork for great adoption.

The best way to get a group of people together to understand business processes is through workshops.

Key items to keep in mind to facilitate effective workshops in a Salesforce project are as follows:

- **Workshop goal**: Be sure to define the goal of the workshop and share this with the workshop participants upfront.

- **Participants**: Ensure you include **subject matter experts** (SMEs) in the workshop. If you have had any indication of business processes varying by region or country, make sure you invite an SME from multiple different regions or countries as far as is practically possible and make sure to align and validate your findings with remaining regions afterward.

- **Workshop environment**: Ideally, you should try to get people together physically in the same room. If not, it can also be done virtually.

  - If physical, make sure the room(s) available are *not* set up in a classic desk structure, but in a manner that encourages moving around and interacting. Prepare brown paper bags or other paper you can plaster on a wall, and bring large post-its and pens.

  - If virtual, make sure you and the workshop participants have access to adequate online collaboration tools.

- **Workshop invitation**: Strive to invite attendees at least 2 weeks in advance and share any preparation you expect the participants to have done before the workshop.

- **Workshop agenda**: A lean workshop agenda will keep focus. A workshop agenda should include the following:

- **Welcome and introduction**: The facilitator welcomes the participants, introduces the agenda and goal, and the participants introduce themselves.

- **Workshop goal**: The goal of the workshop should be shared upfront to make sure the participants understand when the facilitator(s) may interrupt a discussion that is steering off course. Be transparent here to build trust; you are seeking the expert knowledge of the participants to be able to shape the CRM system of your company's future. Share the vision statement you crafted in *Chapter 1, Creating a Vision for Your Salesforce Project*.

- **Parking lot**: This is a common and great workshop technique. Let participants know upfront that any topics not related to the goal of the workshop will be captured in the parking lot for review at the end of the workshop.

- **Ground rules**: Ask the workshop participants to suggest ground rules for how to interact during the workshop. Share an initial suggestion, such as "all participants should feel empowered to be honest about their experiences."

- **Icebreaker(s)**: Set aside time for icebreakers to get the group comfortable sharing and interacting.

- **Breaks**: Set aside time for short 10-minute breaks every 45-60 minutes, depending on the nature of the exercise agenda. If you have just had break-out sessions of 30 minutes followed by 30 minutes of group sessions and presentations, you can opt for 60 minutes between breaks. If your workshop is virtual, be sure to stick to 45-minute breaks.

- **Exercises**: Divide the overall goal into smaller exercises. For example, use the first exercise to determine the current **mid-level process** and identify the **actors** involved. Next, identify what systems and data are used in each process step. This first part could – if effectively led – be done over the course of a few hours, including breaks. Then, spend the second part of the workshop gathering input for what could be improved, what process steps are particularly cumbersome, and what ideas the participants have.

- **Workshop closure**: Round the workshop off by thanking the participants for their input, recap the output of the workshop, and review any items captured for the parking lot.

- **Workshop roles**: A good workshop has a few key roles, some of which may be the same person:

  - **Lead facilitator**: Leads the participants through the exercises, mediates if individuals disagree, and keeps the group moving forward. If you will not be facilitating the workshop yourself, make sure an experienced lead workshop facilitator is assigned and thoroughly briefed beforehand.

  - **Co-facilitator**: Supports the lead facilitator. It's great to divide the facilitation work if multiple sub-groups need to work in parallel on exercises.

  - **Scribe**: Someone should take notes throughout the workshop – at least in the group sessions. If you have two people facilitating, one could be a scribe if you are not able to have a dedicated scribe and timekeeper resource.

- **Timekeeper**: Make sure you stay on track and are able to complete the planned agenda by assigning a timekeeper to state when there are 15, 10, and 2 minutes left in each exercise.

- **Workshop products**: The output of the workshop are the workshop products. These are the gems of knowledge you have been seeking to uncover by having the workshop. Here is a brief description of key workshop products you should seek to produce:

  - **Actors**: Identified actors (end user groups) involved in the current business process.

  - **Processes**: These are the process maps created in the workshop along with various overlays – see the following points:

    - **Variations**: *Pay extremely close attention* throughout the workshop for any process variations identified. For each variation, identify which of the following applies:

    - The variation is *consistent* across all business lines or geographies. Note down these variations as **type-1 variations**.

    - The variation is *inconsistent* across business lines and geographies. Note down these variations as **type-2 variations**.

  - **Systems**: The list of systems the actors work in throughout the process.

    - **Variations**: Pay the same attention as for process variations to any *system* variations identified. Follow the same notation for variations identified.

  - **Data**: What key information/attributes are used by the actors in the process. Keep this at a business-level description at this point, not an actual API name. Critically, get an understanding of data volume. This simply needs to be at *orders of magnitude* (10, 100, 1,000), not specifics.

- **Issues** and **suggestions**: List of pain points and new ideas for improving the process to better serve customers.

> **Important note!**
> Create and maintain a document of the *process issues* and *suggestions* for improvements. You will need this document several times over the course of the entire Salesforce project.

Let's have a look at what process flow chart **PME's Salesforce taskforce** managed to create together with their workshop participants and what insights about the process were uncovered:

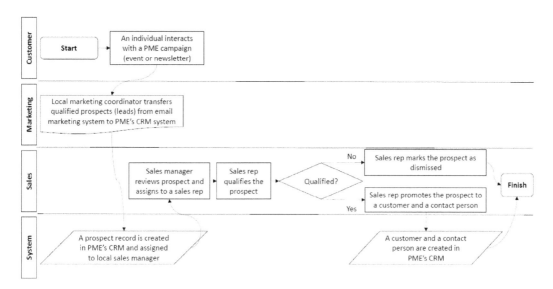

Figure 2.4 – Swim lanes example for PME's current lead management process

The lead management workshop was a remarkable success. The SMEs interacted well and eagerly shared how they work (their processes), in which system they performed the work, and what information (data) they relied on and created.

Kalinda, the business owner in PME's Salesforce taskforce, noted down the variations mentioned in the workshop. Here's what she captured:

- *Process variations:*

  - *Some countries have a business rule where specific information has to be captured by marketing before they can input it into PME's CRM system and assign the prospect to sales. However, this is not supported by the CRM system, but sales do complain if the information isn't there. Other countries don't have any business rules for what information there has to be about the prospects.*

  - *In some countries, it is the responsibility of the marketing department to contact and qualify prospects and update information in PME's CRM system. In most countries, it is the sales team's responsibility.*

- *System variations:*

  - *Many different email marketing systems are used by the local PME country organizations – even systems not approved by PME's IT team.*

  - *Some countries use the CRM system provided by PME; others use it to a lesser extent while concurrently maintaining other data outside the CRM in a spreadsheet.*

The workshop participants gladly shared issues and suggestions for improvements and innovative ideas. Kalinda noted down all these points carefully:

- *Issues raised by workshop participants:*

  - *It is a cumbersome, manual process for the marketing coordinator to transfer leads – errors do happen, and that means wrong customer information is loaded, or worse yet, prospects do not get loaded into the CRM system, and the potential sales are lost.*

  - *PME's current CRM system does not have any duplication checks, so many duplicate prospect records exist in the current system, making it hard to get an overview, and the sales team's trust in the system is generally low.*

  - *Many prospect records are incredibly old and not qualified by sales. In the workshop, sales explained they were not all aware (not trained) on how to convert prospects to customers and contact persons; instead – when qualified – the sales rep simply creates a customer and contact person directly without converting the prospect.*

  - *Sales raised the issue that it often happens that a prospect is assigned by the sales manager to the wrong sales rep – a sales rep without the specific product insights or in the wrong geography. They then have to reassign them back to the sales manager, who reassigns them to another sales rep.*

  - *Sales complained about the quality of prospects assigned to them.*

- *Suggestions from workshop participants:*

  - *Marketing would like it to be possible for the current email marketing system to be able to transfer marketing qualified prospects to PME's CRM system automatically*

  - *Sales would like a few key bits of information to be present on the prospect before it is assigned to them in order to effectively be able to qualify the prospect*

  - *Sales would like it to be a system rule that all prospects must have at least a company name, name, email address or phone number, and product interest filled out before it is assigned to them*

  - *Sales would like to have all old prospects deleted*

  - *Sales suggests training marketing on how sales is organized, for them to be able to correctly assign prospects to them*

  - *Sales would like to have better training on how to convert prospects*

Tip!

Have regular sessions with your executive sponsor to report on the progress and initial findings from the workshops. Share process and system variations identified and ask the executive sponsor whether it is the ambition and intention of the Salesforce project to also carry out **process harmonization** among your company's business lines and local organizations.

Having analyzed and gained a better understanding of your company's current business capabilities, it's time for the next step in determining the capabilities of your Salesforce solution.

Imagining new business capabilities supported by your Salesforce solution

Throughout the activities you have carried out in the Salesforce taskforce up to this point, you have gathered valuable suggestions from the SME end users for improvements to be more efficient and to better serve your customers and ease the buying experience of your customers.

Bear in mind, while end users are one of the most valuable sources of input on what to improve, end users' **horizon of possibilities** will likely span only the edges of the existing systems and processes.

As members of the Salesforce taskforce, it is *your responsibility and prerogative* to look beyond minor incremental process improvements and reimagine what capabilities your company's future Salesforce solution should support in order to innovate the ways your company improves its competitive advantages.

Set up a workshop with *key* stakeholders – including the senior leadership of your company – to understand the capabilities they would like to be part of the future solution.

Prior to the workshop, prepare the following two key items and share them with the workshop participants:

- Create an **executive summary** of the workshops you have carried out – highlight the issues and suggestions shared by the SMEs in the workshops
- Research what **trends** are happening in and beyond your industry and try to understand how your competitors are innovating how they market to and serve customers

The goal of the workshop should be to provide you with a list of desired new capabilities for your Salesforce solution.

PME's Salesforce taskforce has identified a number of new desired capabilities and enhancements to existing business capabilities, which are shown in the following table:

Domain	New business capability	Description
Marketing	Account-based marketing	Sales and marketing would like to be able to target specific accounts with account-specific content either in connection with a lead generation campaign or while a sales opportunity is in progress.
	Customer journey personalization	Marketing would like to create different lead and onboarding journeys for different customer segments, roles, and geographies.
	Consent and preference management	All stakeholders agree that PME requires a solid, transparent, comprehensive, and flexible solution to manage and document consent and preference data.

Sales	Sales forecasting management	Sales want the future Salesforce solution to provide forecasting abilities.
	Quote management	Sales want a solution where they can create a quote (PDF) in the system without having to use Excel and Word.
	Contract management	Sales would like better system support for managing the entire contract lifecycle.
	E-commerce store	Sales management would like to offer customers the option of buying from an e-commerce store in addition to placing orders via phone and email.
	Distribution partner management	Sales would like to have better insights into their distribution partners' sales opportunity pipeline as well as a platform to unify customer issue resolution.
Customer service	**Service-level agreement (SLA) management**	Customer service would like the future Salesforce solution to provide better tracking and management of how well PME performs against their own SLA targets as well as the SLAs contractually agreed with their customers.
	Knowledge management	Customer service – and the wider PME organization – possesses a great deal of knowledge and assets on how their products work and how to troubleshoot and fix various issues. They want this information to be used in a structured manner when the customer service agent works on a customer issue.
	Self-service customer management	Customer service would like to offer customers the option of logging a case and finding help through an online portal – in addition to getting help via phone and email. They imagine the assets described in the previous knowledge management capability could be shown in the portal.
	Field service management	Customer service has a sub-department of field service technicians whose job is to ensure PME's products are professionally installed at customers' sites. PME would like a system for managing the entire field service process.

All	Reporting and analytics	All stakeholders agree that PME requires comprehensive, usable, and flexible reporting and analytics solutions – to have a 360-degree view of customer interactions and to leverage the consolidation of customer data for AI and ML prediction and prescription to anticipate and respond to customer demand.

Table 2.1 – Desired new capabilities for PME's Salesforce solution

Some of the key findings of PME's Salesforce taskforce were an e-commerce store as a new business capability with PME's sales domain and field service management as a new capability to be system-supported by the new Salesforce solution.

At this stage, you are busy gathering information, creating process maps, understanding your company's current business capabilities, and imagining new capabilities to be supported by your future Salesforce solution. As you progress through this stage, let's look at how a **RAID** log will help you in your process.

Starting to build your RAID log

At this stage, you will already have encountered many different stakeholders with various inputs, suggestions, and expectations of the project.

Be sure to capture RAID items related to the project in a repository available to the Salesforce taskforce team and key stakeholders. RAID stands for the following:

- **Risks**: Log any risks you may have uncovered as you progress through the phases of your Salesforce implementation. Capture any actions you are making to prevent or mitigate the risk.

  - **Probability-impact matrix**: Beyond a risk title and description, be sure to note down what the impact would be if the risk were to materialize, and what the probability is for the risk happening. This categorization helps you prioritize risks and focus on the prevention or mitigation of high-impact and high-probability risks.

- **Assumptions**: For any assumptions you make throughout your Salesforce project, make sure to capture them. Add categories for easy sorting.

- **Issues**: If a risk that has already occurred is having an impact, this is now an issue to be handled. Keep a log of ongoing issues, including actions you are taking to fix or mitigate them.

- **Decisions**: Capture any decisions made by the Salesforce taskforce, independently or in agreement with your executive sponsor and wider leadership team. Be mindful that things are not yet set in stone. As such, decisions made in the first iteration of the pre-development phase may be modified as you capture more insights and gain more knowledge. Be open to change at this stage – the nature of your Salesforce project is being shaped to better fit with your organization's business needs. That's a key factor for ensuring the success of your Salesforce project!

Now that you have gathered all the business capabilities for your Salesforce solution, it is time to look at the important technical enablers for your solution to be able to function.

Determining the required technical enablers to support the identified business capabilities

You have spent a great deal of effort understanding the business processes and desired new capabilities. What you might not have focused a lot on is the effort to understand and describe the **technical enablers** and **non-functional requirements** for your Salesforce solution.

Let's start by understanding technical enablers.

Technical enablers and non-functional requirements

These requirements are critical capabilities to ensure the Salesforce solution will be performant and scalable, offer a seamless user experience, and maximize adoption.

Let's look at some of the most common technical enablers and non-functional requirements for Salesforce solutions:

- **Localization**:

  - **Multi-language support**: For any company operating in more than one country, or in a country where there is the use of, or it is standard to offer more than one language for users of a system, multi-language is often a requirement.

  - **Multi-currency support**: Similar to the multi-language support requirement, for any company operating in multiple countries with multiple currencies in use, having a solution with multi-currency support is often a requirement.

- **Identity and access management (IAM)**: As a company, you need to ensure that the right people (and systems) have access to the right features in the system, and you want to make sure only the users authorized for that access can, in fact, access your systems and features. This is where IAM comes in. Two of the most known use cases are shown in the proceeding list. You want to ensure security and ease of access for authorized persons:

- **Single sign-on (SSO)**: SSO allows a person to only remember one set of credentials (username and password) to log in to multiple systems. It is commonly used for enterprise systems, where internal users log in to internal applications using SSO for seamless access for authorized users. The **provisioning** and management of user access is a significant part of IAM.

- Another implementation use case for SSO is called **social sign-on** where external users use their social media accounts (for example, Facebook, Twitter, and LinkedIn) or other commonly used services (for example, Gmail and Amazon) to access digital services.

- **Data security**: When your system identifies (*authenticates*) a user, your system should only provide that user with access to data that the user needs to execute their work. Data security is concerned with ensuring the right level of data access to the right users at the right time.

- **Data and file archival**: As you migrate data from legacy systems, and your company users begin to work in your Salesforce org, data will be created, and lots of it. You will need and want to have the ability to archive data for several reasons:

  - **Regulatory requirements**: Many regions, countries, and states only permit **personally identifiable information** (**PII**) data retention for a certain period, and only if there is a valid reason for the retention.

  - **Usability**: Without regular archival of old or irrelevant data, users will unnecessarily have to sift through all the data clutter, decreasing the user experience and productivity, and increasing the risk of errors and poor customer experiences.

  - **Performance**: If you don't archive old or irrelevant data, when your users search for information, load record pages, or run reports, the load time will increase over time, diminishing the user experience and productivity.

  - **Cost**: Storage is a big cost driver for cloud solutions. While you can purchase additional storage, having an *efficient and automatic* archival solution in place will quickly outweigh the saved storage space cost.

- **Backup and restore**: Whereas data and file archival is mostly relevant for old or irrelevant data, backup and restore is concerned with ensuring business continuity by making sure your critical operational Salesforce data is backed up regularly and easily restorable should data be unintentionally deleted, modified, or corrupted for any reason. There are robust Salesforce-specific **commercial off-the-shelf** (**COTS**) solutions available that provide these services – **OwnBackup** and **Odaseva** are the market-leading providers in the Salesforce ecosystem.

- **Accessibility management**: Offering a Salesforce solution that is accessible for people living with some form of disability is not only ethical but also smart for business. In many countries, it is a legal requirement that goods and services offered to customers must be accessible also to people with disabilities. Furthermore, being able to offer employment to people with disabilities will increase your talent pool – to the benefit of the people with disabilities as well as your company.

- **Speed/performance**: Everyone wants solutions that are quick and easy to use with no wait time. In some use cases – for example, a customer service call – having a customer contact page load time exceed a certain number of seconds may not be acceptable to the business, as it would result in a poor customer experience. Other key performance aspects are running reports and dashboards and searching for records. These are examples of non-functional requirements.

- **Reliability**: The percentage uptime of the system where users are able to log in and perform work is a characteristic of reliability. Although your company will be largely depending on Salesforce to ensure uptime, there are things you can do to minimize downtime. These include minimizing interruption of operations when deploying new features and building in exception handling to gracefully handle connection issues for your integrations.

Many non-functional requirements are effectively provided and, to some extent, controlled by Salesforce since – in addition to being a SaaS solution – it provides infrastructure and a platform for its customers. However, how you design, configure, and develop your Salesforce org has a major impact on its level of security and performance.

> **Tip!**
> Engage a technical architect to support you in defining the technical enablers for your Salesforce project.

Having covered some of the common technical enablers and non-functional requirements for Salesforce solutions, let's look at data sources that are often integrated with Salesforce.

Data sources

While your company's ambition and intention may be for Salesforce to be the main system for your marketing, sales, and customer service teams, there will inevitably be other systems and applications in your company's **system landscape**. These other systems and applications serve their own unique purpose and hold data that is sometimes relevant for users in Salesforce, while other times it may be Salesforce that needs to send information to an application to complete a process.

Let's look at some systems that are often integrated with Salesforce:

- **Enterprise resource planning (ERP) system**: These systems are used for operational purposes, covering production planning, stock management, logistics planning, order processing, and invoicing.

- **Data warehouse (DWH)**: This is a database containing data from across the enterprise. It may have various levels of distribution and structure. For example, data lakes contain unstructured data for data scientists to model and leverage, whereas a highly structured and managed database can provide the basis for BI systems to provide a reporting layer to users.

- **Active Directory (AD)**: An AD often serves as an **identity provider** or **identity store** used in relation to *IAM* to centrally manage provisioning and **deprovisioning** and access level changes.

- **Website(s)**: Your company likely has at least one website describing your company and its products and services. The website may have a **contact form** for people to fill out with questions or to request information, which is then sent to your sales or customer service teams.

- **Content management system (CMS) and digital asset management (DAM) system**: These systems store assets used across a number of applications, including your website and email marketing solutions.

- **Product information management (PIM) system**: Stores information about your products (and services) used across a number of applications, including ERP, e-commerce stores, CRM, and product **configuration, pricing, and quoting (CPQ)** solutions.

> Tip!
> Review your workshop products of *systems* and *data* as a baseline for integrations with your Salesforce solution. Engage an architect to support you in determining which systems and data (external to Salesforce) need to be integrated with your Salesforce solution.

Next, we'll look at how to summarize all the capabilities of your Salesforce solution.

Summarizing the capabilities to be supported by your Salesforce solution

Having added the technical enablers, you now have a complete set of capabilities in scope for your Salesforce project. Let's see how the business capability map looks for PME in the following figure:

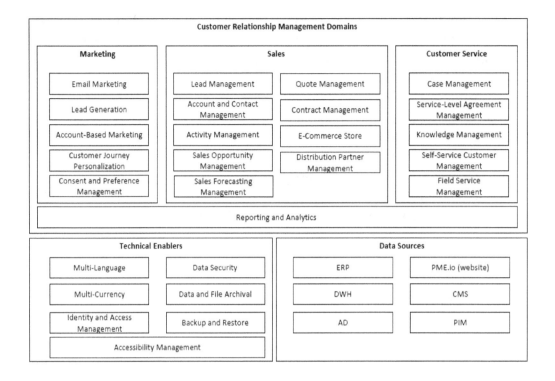

Figure 2.5 – Business capability map for PME's Salesforce solution

In addition to the imagined new business capabilities, PME – being a global company – identified the need to have a localized solution offered in multiple languages and supporting multiple currencies. PME is also keen on ensuring solid security, both in terms of user access to and in Salesforce, as well as what data they can access. PME also sees the value in regularly archiving and backing up data and files to comply with regulations and making sure the solution is scalable and performant, and ensuring business continuity. Finally, PME wants to ensure their solution can be used by their employees with disabilities – as well as by people with disabilities working for their customers.

PME identified – through workshops with end users as well as engaging with a technical architect – what key systems and data sources are required to work in conjunction with the future Salesforce solution.

Having summarized the capabilities to be supported by your Salesforce solution, you are ready to move on to determining the high-level technical scope required for your Salesforce solution.

Determining the high-level technical scope of your Salesforce solution

Until now, we have not spent a lot of energy on Salesforce features. That will change now as we progress to define the high-level technical scope of your Salesforce solution.

You have defined the business capabilities to be supported by your Salesforce solution. Now it's time to understand what **hypothetical technical scope** is needed to support the business capabilities. It is *hypothetical* at this stage because you will undoubtedly have many assumptions and uncertainties. As you progress through the chapters in the pre-development phase, the technical scope will become firmer until you finally contract with Salesforce.

The high-level technical scope

At this stage, determining the high-level hypothetical technical scope serves the following purposes:

- Being able to estimate the **license cost** of the solution

- Being able to request one or more potential implementation partners to provide an estimate for the **implementation cost** of the solution

- Being able to estimate the *cost of the* **target operating model** to maintain and run your Salesforce org

- Being able to understand the technical impact and reach on your company's overall system landscape

All of these purposes are key for the proceeding activities in the pre-development phase, business case modeling, as well as being a key part of an RFP document should you choose to engage with an implementation partner to take part in the delivery of your project. More on that in *Chapter 4, Securing Funding and Engaging with Salesforce and Implementation Partners.*

Let's start by understanding the different Salesforce license types.

Salesforce license types

The first concept to grasp is the most wide-reaching and fundamental – let's take a look at Salesforce org editions.

Org editions

Salesforce *core* comes in four different **editions** for commercial use, with a fifth edition, **Developer**, for non-commercial use. For simplicity, let's focus on the four editions for commercial use as described at the following link: `https://help.salesforce.com/s/articleView?id=sf.overview_edition.htm&type=5`.

The edition of your Salesforce org determines what features and services you have access to and is also the baseline for the cost of your **user licenses**.

You will want to study the differences between the editions in detail and weigh them against your company's requirements. If you're planning to use Salesforce to support both sales and customer service, study the *Compare editions and top features* section at the following page: `https://www.salesforce.com/editions-pricing/sales-and-service-cloud`.

The following are a few key points to consider when deciding on which edition to go for:

- You can upgrade your edition, but you *cannot downgrade it*. As such, if you are on the fence about which edition to choose – as you may not yet understand exactly what the different features mean – opt for the lower edition.

- If you are a medium or large company and planning to have multiple **actors**, use Salesforce across multiple business domains, automate your business processes, and integrate with other systems, you should consider choosing either the *Enterprise* or *Unlimited* edition.

> **Tip!**
> If you are unsure which edition to choose and have decided – at this stage – to assume a lower edition will suffice, add this to the assumption section in your RAID log as well as to the risk section with a *high-impact* categorization as the cost of upgrading will have a great financial impact.

Let's move on to explore Salesforce clouds.

Salesforce clouds and add-ons

Salesforce (the company) offers a vast variety of products and features bundled by domain, called **clouds**. In *Figure 2.6*, you'll see how Salesforce illustrates its products, collectively making up the brand name **Salesforce Customer 360**.

Salesforce products range from *core CRM* products, such as **Sales Cloud**, **Service Cloud**, and **Experience Cloud**, to **marketing**, **commerce**, **integrations**, and **analytics**.

Sales Cloud and Service Cloud are priced on a per-user basis whereas the other clouds have varying cost models. For example, for Marketing Cloud, the cost model is on a per-contact basis, whereas for Commerce Cloud, it is based per order.

To extend core Salesforce, there is an abundance of **add-ons** for almost every conceivable use case – and it is growing every year. You can view all the available add-ons in the document displayed at the following link: `https://www.salesforce.com/content/dam/web/en_us/www/documents/pricing/all-add-ons.pdf`.

A key point about add-ons is that they are not all available with all editions, meaning *some add-ons require a higher edition* to be able to purchase them. For example, in order to purchase and use **Sales**

Cloud Einstein for AI-powered sales intelligence, **Partner Relationship Management** (**PRM**) for building a community site to collaborate with partners, or **Field Service** to manage and support your customers with on-site technicians, you need to be on either **Enterprise** or **Unlimited** org edition.

Next, we'll explore how AppExchange can further support your Salesforce solution.

AppExchange

To extend Salesforce clouds' common use cases, **independent software vendors** (**ISVs**) develop and offer Salesforce **apps**, pre-built solutions, and templated Salesforce **Flows** – to accelerate the value delivery of your Salesforce project.

AppExchange apps also include connectors for the most popular and widely used enterprise apps, such as **LinkedIn**, **SurveyMonkey**, **Slack**, and **CTI** solutions, to name but a *few*. Salesforce also has its own department called **Salesforce Labs** offering free AppExchange apps and solutions with a selection shown at the following link, where you'll see an overview of AppExchange solutions specific to Salesforce clouds `https://appexchange.salesforce.com`.

While some AppExchange products are paid for and others are free, you should be aware that none of them are legally or technically offered by Salesforce Ltd. Therefore, before including any AppExchange product in your (Salesforce) system landscape, be sure to carry out the same **due diligence** you would in relation to any other SaaS product purchase.

Furthermore, also listed on AppExchange is a menu dedicated to official Salesforce *consultants*, also known as **system integrators** (**SIs**) or **implementation partners**, which is the term we are using in this book. We'll return to explore this section further in *Chapter 4, Securing Funding and Engaging with Salesforce and Implementation Partners*.

Crafting your hypothetical high-level system landscape

Having determined the capabilities to be supported by your Salesforce solution, and been introduced to Salesforce products, it's time to craft your hypothetical high-level system landscape. But where do you start?

Start by determining which systems and applications will be part of the Salesforce realm – products within the Salesforce org, add-ons, and any AppExchange apps you may include – and which are *external* to Salesforce.

If you're planning to replace any legacy CRM systems with Salesforce, the capabilities previously supported by those systems will in the future be supported by Salesforce solutions. Determine which Salesforce product covers your required business capability.

For all other systems and applications, refer to the executive summary of your workshop products – the list of systems and data currently used. This will typically include your ERP, DWH, and potentially a **middleware** solution if you're not considering replacing this as part of your Salesforce implementation.

Let's see what hypothetical high-level system landscape the PME Salesforce taskforce has crafted in the following diagram:

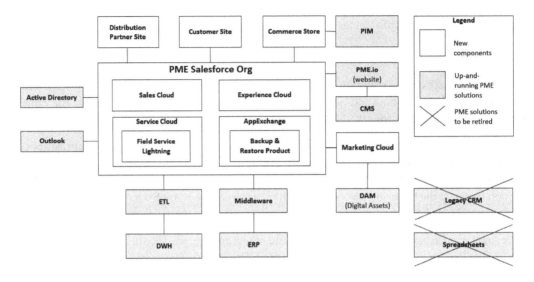

Figure 2.6 – High-level technical scope of PME's Salesforce solution

*Cary Sharp, the PME Salesforce taskforce's **tech owner**, took charge of researching and leading the effort of determining PME's hypothetical technical scope for PME's Salesforce solution to support the required business capabilities.*

First off, Cary reviewed available Salesforce org editions and is currently unsure whether to opt for Enterprise or Unlimited. He has assumed the Enterprise edition will suffice at least until the first roll-out.

*Let's review the **new components** Cary has identified for PME's system landscape:*

- *Salesforce components:*

  - **Sales Cloud**: *To support the capabilities in the sales domain for PME, Sales Cloud is the fundamental Salesforce product.*

  - **Service Cloud**: *To support the capabilities in the service domain for PME, Service Cloud is the fundamental Salesforce product.*

    - **Field Service cloud add-on**: *To support the specific field service management capability, Cary Sharp found out Salesforce has a specialty product for field service use cases and added it to the system landscape.*

  - **Experience Cloud**: *Cary was thrilled to find out Salesforce has a solution for creating customer-facing solutions, so Experience Cloud was added to support two separate capabilities.*

- **Distribution partner site**: *A site where PME's distribution partners can collaborate with PME on sales opportunities and on solving customer cases. PME will have the added benefit of increased visibility of their distribution partners' performance to further support and develop the relationships.*

- **Customer site**: *A site where customers can access PME knowledge resources and request help through case creation.*

- **Commerce Cloud**: *Cary was impressed to see a commerce product available within the Salesforce product portfolio. The commerce site will support PME's ambition of letting their customers place orders themselves online without having to call or email PME's busy sales and customer service departments. Commerce Cloud will also connect to PME's existing* **PIM** *system as the source of product data.*

- **Marketing Cloud**: *To support PME's marketing ambitions, Marketing Cloud is added to the target system landscape. Cary notices there are two different Marketing Cloud solutions to choose from but isn't sure which one to choose. For now, he notes down Marketing Cloud Account Engagement as an assumption as it is mentioned to be ideal for B2B companies. Marketing Cloud will also connect to PME's existing* **DAM** *solution for digital asset management for digital campaigns.*

- **AppExchange**: *To support the* **backup and restore** *capability, Cary notices that these products will connect to PME's Salesforce org through an AppExchange app.*

Next, let's review the **existing components** *to be reused in PME's system landscape:*

- **AD**: *To manage the provisioning, deprovisioning, and access control of the Salesforce users of PME's Salesforce org, Cary recommends reusing PME's existing AD.*

- **Outlook**: *While PME's ambition, in general, is to reduce the use of email, PME understands that email is still a key communication tool – especially for sales reps to engage with their customers. Cary wants to let sales reps sync their Outlook contacts and calendars to support the activity management capability.*

- **ETL**: *PME has an existing ETL that is currently connecting their existing on-prem CRM system to their* **DWH**, *which is used for data archival and aggregated BI reporting. Therefore, Cary wants to reuse the ETL and connect it to their Salesforce org. Cary is unsure whether PME's old ETL solution is suitable for connecting to a cloud-based CRM, so he adds this to the RAID log's assumption section.*

- **Middleware**: *PME has an existing middleware solution that connects their backend* **ERP** *system and current CRM system. Cary's assumption is that the current middleware is fit for modern enterprise architecture. He adds this to the RAID log's assumption section.*

- **Website**: *PME wants to continue collecting leads through their current website,* www.PME.io, *and have the leads sent to their Salesforce org for processing by sales. Their website relies on a CMS for content and localization management.*

Finally, the systems that won't be part of PME's future system landscape are as follows:

- **Legacy CRM**: *The system that is currently being provided for PME's local countries to support their current CRM processes and data.*

- **Spreadsheets**: *Cary recalls all too well the findings from workshops, where many end users mentioned working on and maintaining customer and activity data offline in spreadsheets because their current CRM system does not fulfill the needs they have. PME's Salesforce taskforce all agree it should be a key goal for their Salesforce project to remove the use of spreadsheets to maintain customer data.*

In the case of PME, recall they identified some process variations, but since these variations were consistent across all local countries, Cary assumes all of PME can work within one Salesforce org. He adds this to the RAID log's assumption section.

> **Important note!**
>
> If your workshops and analysis at this stage have revealed *significant* variations in business processes, reporting requirements, or data segregation requirements, you need to carry out an important analysis to determine your **org strategy**. Org strategy concerns determining whether your company will require one or multiple Salesforce orgs to support your company's requirements. Determining an org strategy is beyond the scope of this book. You will need to engage with an experienced Salesforce technical architect to assist you in this analysis.

The following are some key points to keep in mind about the current state of your hypothetical technical scope for your Salesforce solution:

- It will be further modified, refined, and enriched in detail as you progress through the pre-development phase and beyond.

- It is *abridged* for Salesforce, meaning numerous other systems may well exist in your company's **system landscape**. As such, your company's complete system landscape will have many more components than shown in the simple example for PME.

On that note, let's see how you may iterate over defining the nature of your Salesforce project.

Iterating in the pre-development phase

Scoping a Salesforce project is a big and interesting exercise – one that always takes more than one iteration. As you progress through the following activities of the pre-development phase, you will return to review the capabilities to be supported by your Salesforce solution and the high-level technical scope. As you do, make sure you update your RAID log to capture decisions and assumptions. You are well on your way to making your Salesforce project take shape.

Summary

You have now established a hypothesis of what the business and technical scope should be for your Salesforce project – the nature of your Salesforce project. You started by learning how to gather input and determine the desired business capabilities. Then, you were introduced to the Salesforce product portfolio. Finally, you had a first attempt at defining the high-level technical scope of your Salesforce solution.

So far, you have created a vision for your Salesforce project – the *why* – as well as the nature of it – the *what* – and you are ready to move on to the next chapter of your Salesforce implementation: *Chapter 3, Determining How to Deliver Your Salesforce Project*. It is time to look at the best way to deliver your Salesforce project – the *how*.

3

Determining How to Deliver Your Salesforce Project

In this chapter, you will learn about different **delivery methodologies**, and how to choose the delivery methodology most appropriate to your Salesforce project considering the nature of your project and your company's circumstances and priorities. Next, you will come to understand why a well-thought-out **change management** strategy is critical to the success of your Salesforce project. Finally, you will learn to envision the structure of your **Salesforce Center of Excellence (CoE)** – the management and governance framework to oversee your Salesforce program.

You will be introduced to **waterfall**, **agile**, and **hybrid agile** delivery methodologies, as well as the concepts of different project phasing approaches. You will get an understanding of how to craft a communication and change management strategy, along with the related activities required in the different phases of your project. Finally, you will be introduced to the roles and responsibilities in your Salesforce CoE, drafting its CoE **charter**, and how to create a **roadmap** for your Salesforce program.

We will continue to follow our scenario company, Packt Manufacturing Equipment (PME), and their progress through the activities to determine how to deliver their Salesforce project.

After determining how to deliver your project, you'll be ready to move on to the next activities in the pre-development phase of your Salesforce implementation.

This chapter will cover the following main topics:

- Choosing your delivery methodology
- Determining your change management strategy
- Envisioning your Salesforce Center of Excellence
- Iterating in the pre-development phase

Let's go!

Choosing your delivery methodology

In the previous chapter, you defined the nature of your Salesforce project – the desired business capabilities, and the high-level hypothetical technical scope. As illustrated in the following figure, it is now time to determine how to deliver your Salesforce project.

Figure 3.1 – Activities in the pre-development phase

Before jumping to delivery methodologies, let's briefly look at what is special about enterprise projects, and Salesforce projects in particular.

Enterprise projects

Considering all the types of projects an organization may be planning or currently delivering, it is important to note the differences between them.

The following table provides a simple overview of the nuances between some of the most common enterprise project types:

Type of project	Example	Develop in production	Involve backend systems	Customer-facing	Mission critical for business continuity
Business Projects	*Strategy projects, HR, and re-organization projects.*	N/A	No	Perhaps	No

Marketing Microsite	*Website for lead generation for new product launches.*	Sometimes	No	Yes	No
ERP Projects	*SAP or other ERP implementation to support back-office business operations.*	Never	Yes	No	Yes
CRM Projects	*Span simple applications for basic account and contact management to enterprise-grade CRM implementations.*	Sometimes	Sometimes	Sometimes	Sometimes
Salesforce Implementation Projects	*Span basic usage of core functionality to full-fledged implementations with customer-facing experience cloud sites, marketing automation, advanced analytics, and integrations with backend systems.*	Should never happen	Most often	Sometimes	Most often, and increasingly common

Table 3.1 – Enterprise project types

A key takeaway from comparing the types of enterprise projects in *Table 3.1* is regarding **development in production**. For business projects, this is simply not relevant as they are not IT projects. For the remaining project types, it may be the case that projects and solutions are developed directly in a production environment, simply because there are no supporting development environments for the system, or because the risk or the impact of the risk is considered low since the application is not mission critical for the company.

> **Important note!**
>
> For **enterprise-grade** Salesforce projects with solutions integrated into multiple systems and with mission-critical processes supported by Salesforce, *development should never be directly in production*.

Let's move on to Salesforce projects and programs.

Distinguishing between Salesforce projects and Salesforce programs

It is important that you determine whether you're working on a Salesforce program or a Salesforce project. To help understand the distinction, let's explore the following points:

Salesforce project:

- One or more capabilities to deliver
- Has a target deadline

Salesforce program:

- Contains a **portfolio** of Salesforce projects in various phases and/or a number of agile product development teams – a concept we'll explore further in *Chapter 12, Evolving Your Salesforce Org and DevOps Capabilities*
- Does not have a set end date: the goal of a program should be to continuously improve the business domain capabilities of an organization
- For *one* organization, there should only be *one* Salesforce program

The concept of projects is more relevant to certain types of **delivery methodologies**, which we'll come back to later in this chapter. But first, let's get on the same page by going through common **terminology**.

Common terminology used in Salesforce projects

Misconceptions and wrongly used concepts are widespread throughout the ecosystem. It is a major source of confusion, both for business and technical stakeholders alike.

Let's look at some of the most commonly used terms in Salesforce projects:

- **Minimum Viable Product (MVP)**: A term used for product development where a product may be physical or non-physical (digital). In any case, an MVP is a version of a target *product* (solution) with a *minimum* set of **must-have** features while still being a *viable* solution for users to do their job – in *production*. An MVP inherently lacks many features that it **should have** or would be **nice to have**. Salesforce projects are innovation projects for many organizations. Many aspects of future imagined processes are still assumptions at this point. Believing an organization can know up front exactly what features are most important and bring the most value – let alone the detailed requirements of such features – is a risky and suboptimal approach for most Salesforce projects. Instead, defining an MVP and gaining fast, valuable *feedback for improvements*, is often the fastest and best approach.

  - The concept of an MVP can be applied at multiple levels:

    - At the *project level* when determining which collection of business capabilities are the minimum that will make for a viable solution for end users.

- At the *capability level*, where an MVP is about scoping which features are critical and which are not.

- **Pilot**: A **roll-out** approach used to get feedback – often on an MVP version of a solution – from a limited group of the total intended users of a solution. This valuable feedback can then be used to enhance the solution prior to rolling it out to more users.

- **Minimum Marketable Product** (**MMP**): Building on the concept of an MVP, the MMP is a version of a product (Salesforce solution) that is believed to be more likely to be accepted or bought (adopted) by the market (Salesforce users) than the earlier MVP version.

- **Proof of Concept** (**PoC**), **prototype**, and **pretotype**: These terms describe the activity of attempting to create versions to see if it is conceptually and technically possible. The terms (solution versions) are grouped together as they share a key trait: they are never meant to go to production. A PoC – even if it proves a concept (solution) works in a development environment – will always require further refinement, re-engineering, design optimization, re-building, and testing *before* being ready for production.

  - Experienced and less experienced professionals, unfortunately, often inaccurately use these terms. Some use the terms to explain or request a slim version of a feature or project to be delivered fast and used in production without alignment or articulation of the purpose up front, not considering the necessary technical enablers or dependencies to other features, systems, or projects.

- **Solution**: Refers to the technical deliverable of the project, including standard Salesforce out-of-the-box components, as well as any configurations, customizations, and integrations built into the project.

- **Implementation**: A term often used synonymously with the terms project and solution.

- **Release**: This can have two meanings with subtle differences. One meaning of release is when a solution is deployed to production. Another meaning is – if some form of feature flag was implemented together with the feature – when the feature is *turned on* and is ready to be used. We'll cover this concept further in *Chapter 7, Building and Testing Your Initial Release*.

- **Salesforce** or **Salesforce, Inc**: This can either mean the actual system your users will work in or the company *Salesforce, Inc*.

Many of the terms described in the preceding section relate to some form of agile development methodology, and – if you choose to use an implementation partner for the delivery – it should be clearly described in your contract. We will return to contracting in *Chapter 4, Securing Funding and Engaging with Salesforce and Implementation Partners*.

> **Tip!**
>
> A *general rule of thumb* is – if at all possible – you should *prioritize a tighter scope* for an MVP for faster delivery and faster feedback *over* a comprehensive solution that is delivered slower, with more risk of being outdated once it is taken into use.
>
> Once your MVP is piloted with a select user group, you can iterate to improve the MVP into an MMP for further roll-out.

Whether or not agile is the right delivery methodology for your Salesforce project will be covered in the next section.

Understanding delivery methodologies

Let's jump right in by understanding the two extremes of delivery methodologies, and a third that is harder to define but often used.

Pure waterfall

In **pure waterfall** Salesforce projects, requirements are defined up front in full detail by business stakeholders through the help of business analysts. All requirements are nicely organized in the behemoth **business requirements document (BRD)**.

Next, the architects and sometimes senior developers get to work by creating detailed low-level solution designs. The solution design is then handed off to the development team to develop the entire solution. If the solution is an enhancement or extension to an existing, live system, **system integration testing (SIT)** is carried out before **user acceptance testing (UAT)**.

Then, UAT is carried out in strict accordance with the BRD stated acceptance criteria, and the technical, performance, and non-functional requirements are tested as well. If one or more specific acceptance criteria fail, the project leadership needs to scramble to assess whether the failures are critical and must be fixed before deploying or to deploy and fix them in a later release. If the solution passes UAT, the solution is deployed and rolled out to users.

If you have contracted with an implementation partner to deliver the project, the period from architecting and developing the solution against BRD *could take months* before UAT starts and you see what solution has been developed.

> **Tip**
>
> Learn more about waterfall delivery methodologies in these links at *Prince2* (`https://www.prince2.com/eur/what-is-prince2`) and *PMP (Project Management Professional)* (`https://www.pmi.org/certifications/project-management-pmp`).

Pure agile

In **pure agile** Salesforce deliveries, the agile team is armed with nothing but a brief **product vision** as a guide. It is up to the agile team itself – led by the Product Owner – to create their product backlog containing the most important and urgent features to deliver. The team works together to refine the product backlog, and at **sprint planning** the team uses **story point pokering** to estimate the size of the **product backlog items** (**PBIs**) (aka **user stories**). Once the team's sprint capacity (of *user story points*) is filled, the team gets to work to create solution designs for the PBIs, develop and test them, and then **demo** the developed features to stakeholders (a form of UAT) before finally releasing the developed and accepted PBIs. The **sprint** (also called an **iteration**, depending on the specific agile methodology chosen by the team) usually lasts between one to four weeks, with a two-week duration being the most prevalent.

The agile team is **self-organized** in the sense that it will develop solutions for the requirements prioritized and created as PBIs. The PBIs may be organized at various levels, such as **themes**, **epics**, **features**, **user stories**, and **tasks**. The organization of PBIs is not the important part to distinguish agile from waterfall deliveries. The key aspect to understand with pure agile is that the team does not have a set scope defined up front – at most, a product vision is articulated by the funding party, which the PO must translate into a collection of PBIs that *collectively* form the **sprint goals**.

We'll cover these topics and terms in more detail in *Chapter 7, Building and Testing Your Initial Release*.

An agile team *may* contain the following roles, which are more or less formalized depending on the type of agile methodology:

- **Product Owner** (**PO**): Represents the business and the intended users of the solution.

- **Scrum master**: Facilitates the agile ceremonies, removes impediments, coaches the team on the agile methodology, and ensures agreed actions for continually improving the team collaboration are executed. A key responsibility of the scrum master is to shield developers, functional consultants, and QA specialists from disturbances, such as meetings and reporting.

- **Developer** (**coder**): Develops complex logic, customized **user interfaces** (**UIs**), and custom integrations using **Apex** and **JavaScript**, among other technologies.

- **Functional Consultant** (**configurator**): Configures functionality **declaratively** – using clicks, not code. In a Salesforce context where **low-code/no-code** is a distinguishing feature, functional consultants are usually more plentiful in an agile team than developers.

- **Business Analyst**: Some agile teams have dedicated business analysts to support the PO in **backlog refinement**; in other teams, the functional consultant or developer takes up that role.

- **QA Specialist (tester)**: Some agile teams have dedicated QA specialists to carry out user story testing and regression testing; in other teams, the functional consultant and developers share the user story testing responsibility, and the PO is responsible for regression testing – if it is a manual process. We'll cover various forms of testing in *Chapter 7, Building and Testing Your Initial Release*. Needless to say, for mission-critical systems – as Salesforce is often considered by organizations – investing in quality assurance is imperative.

In the following figure, you can see how the two methodologies compare visually:

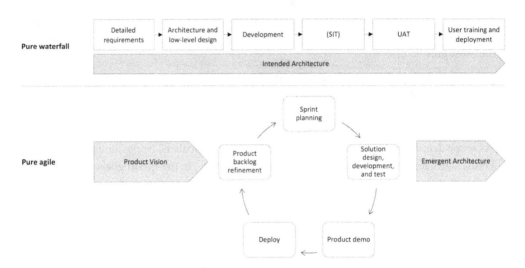

Figure 3.2 – Pure waterfall versus pure agile delivery

Two key points to take notice of in *Figure 3.2* are as follows:

- In the pure waterfall delivery methodology, the requirements are defined in detail up front and the architecture is then designed in detail. The development team then strictly follows the designed architecture, and the initially **intended architecture** is delivered. Whether the intended architecture is the best and most **fit-for-purpose architecture** cannot be concluded until the solution is deployed, and feedback is received from end users. In these **Volatile, Uncertain, Complex, Ambiguous (VUCA)** times, having the audacity – for a large or complex project – to believe you can comprehend detailed requirements and *master-plan* detailed solution designs is a risky approach.

- In the pure agile delivery methodology, the architecture evolves over time and is said to be an **emergent architecture**. A purely emergent Salesforce architecture is rarely the most optimal architecture – at least if zero **guard rails** or **development guidelines** are provided to the agile team.

> **Tip**
>
> Learn more about agile delivery methodologies at the following links:
>
> - **Kanban**: `https://www.atlassian.com/agile/kanban`
> - **Scrum**: `https://www.scrum.org/resources/what-is-scrum`
> - **SAFe (Scaled Agile Framework)**: `https://scaledagile.com/what-is-safe`

As promised at the beginning of this section, there is a third delivery methodology that is harder to define but often used for Salesforce projects.

Hybrid agile

Hybrid agile comes in many forms, as it is not a model defined in a framework like traditionally agile and waterfall delivery methodologies. True-to-the-core agilists might characterize SAFe as a hybrid agile framework, but settling that argument is out of the scope of this book.

In a hybrid agile delivery methodology, you will *pick elements* of each extreme to come to a middle ground that is *better for your organization* than either of the extremes in pure form.

The purpose of hybrid agile delivery is to have some *indication of cost and solution architecture*, but not to the extent that too much time is spent on determining all the lowest-level details up front – at the cost of the requirements becoming stale before the project's development phase can commence.

Choosing your Salesforce delivery methodology

You should consider choosing the *pure waterfall* delivery methodology if the following apply:

- You can accurately articulate your organization's and business users' detailed requirements – in *a Salesforce context* – up front in a BRD.

- You believe providing the requirements, then waiting in the dark for the development period (weeks, months, or even quarters) won't risk the initial requirements becoming obsolete before the solution is delivered.

- You value knowing the final cost and timeline over continuously inspecting the solution throughout development, and being able to adapt to changing requirements, new knowledge, or new uncovered key requirements.

- You want to be sure to avoid **scope creep**.

- Your project's scope is limited, requirements are not complex, and your business stakeholders who provide requirements cannot be available for an extended period of time.

- Your project *solely* involves *either* of the following:

  - Non-negotiable legal or compliance features

  - Complex integrations where independent teams or persons are dependent on each other being available at specific times

- You are prepared to *potentially* pay a higher price for delivery in a **fixed-budget-fixed-scope** model versus a **time and material** model.

On the other hand, you should consider choosing the *pure agile* delivery methodology if all of the following are true:

- You already have a Salesforce org live in production and simply want to add new features *to existing Salesforce clouds*.

- You do not require any estimate of the total cost to deliver your scope.

- You accept the *risk* associated with a *purely emergent architecture* (unless also putting in place suitable governance mechanisms).

- You value continuously inspecting the solution and being able to adapt to changing requirements, new knowledge, or new uncovered key requirements *over* knowing the final cost and timeline.

- You believe you will want and need to do a lot of prototyping PoCs to understand what a solution will entail – *and accept this approach will require time and money.*

- You are aware of and accept the risks associated with a time and material contract model should you engage with an implementation partner to work with you in a pure agile delivery methodology.

You should consider choosing a form of a *hybrid agile* delivery methodology – instead of pure agile – if the following apply:

- Your organization requires *at least some indication* of the cost of your project.

- You see the value in *some level of architecture* being determined up front.

- You need to approach *some* elements of the project scope in a waterfall manner, such as legal requirements, compliance requirements, and integrations while needing to approach *other* elements in an agile way.

In our scenario with PME, the Salesforce Taskforce's project is charged with the goal of developing an MVP solution including capabilities for marketing, sales, and service, and rolling the solution out globally. The solution must have sufficient functionality that it can replace and retire a legacy CRM currently serving sales and customer service departments, and a simple email marketing system.

Project phasing approaches

In the olden days, deploying IT solutions was done in a big bang manner, deploying all functionality and having all users going live at once, celebrating in style with champagne and cake – or so I am told. For at least the past decade, it's become the norm and mantra to build an MVP to release a working solution as soon as possible.

For many types of IT projects, it makes sense to build an MVP and release it as soon as the developed product can deliver value through a better feature set, experience, business outcome, and/or lower cost compared to the existing system being replaced. However, for enterprise IT projects – Salesforce implementation projects fall into this category – there are valid reasons to counter the urge to go live in production (the first go-live) prematurely.

The strongest argument *to not rush* the initial roll-out is that the moment you have users live in production, you will multiply your Salesforce setup and complexity by at least two-fold compared to being in the development phase.

Here are some of the key drivers and responsibilities you will have once you go live that you don't – *at least not to the same degree* – in the development phase:

- User provisioning and deprovisioning:

  - Employees change roles and need new roles, permission sets, groups, and queue memberships

- The training of new users

- Managing the production environment

- Potentially managing a production support environment

- Regression testing key processes three times a year with every Salesforce release

- The handling of new feature requests from users

- The handling of bug reports from users

This can seem like a lot of work, and it is if it's not managed in the leanest way possible, process-wise, and architecturally. We'll cover this further in *Chapter 12, Evolving Your Salesforce Org and DevOps Capabilities*.

I am by no means suggesting you push your initial release for as long as possible. I am simply stating *there is a consideration to be made for the right timing* – factoring in the time it takes to get these processes in place, some of which require technical setup to be done up front.

There are ways of phasing your implementation so as to not create a big bang where you are overwhelmed with the previously mentioned production responsibilities.

Two types of project phasing exist that at first may seem identical, but there are distinct differences.

Phasing by scope

Different business domain capabilities are developed and released to relevant **functional groups of users**. For example, you could opt to develop and release the business capabilities for your *sales* organization and later develop and release for your *marketing* and *customer service* organizations. There are pros and cons to this approach, as highlighted here.

These are the advantages:

- Smaller scope means faster development and roll-out of value-adding, working solutions
- Acquired Salesforce licenses are utilized sooner than if the solution had to include the full scope, and wait for all capabilities to be developed before any single functional user group could start using Salesforce
- Less concurrent effort for training and change management
- Less compounded risk of any one business capability not working and delaying the deployment of the entire solution

These are the disadvantages:

- Risk of getting stuck optimizing a few capabilities beyond the return on investment at the cost of not developing new capabilities for other user groups.
- Potentially require temporary integrations to legacy systems to be developed to have a full customer overview. For example, if a Salesforce solution is initially built for sales users and doesn't include customer service, the sales users won't have access to their customers' service history in Salesforce, and will either have to use two CRM systems (poor user experience), or a temporary integration needs to be developed to display service history from the legacy CRM in Salesforce.

> Tip!
> As we progress through the activities in this chapter, bear in mind the maturity and the strategic importance of your different business capabilities, as they will guide you to prioritize the scope for your initial Salesforce release.

Let's look at another strategy for phasing your Salesforce project.

Phasing by geography

In this type of project phasing, users in one geography will be using all the capabilities developed in the project – but the roll-out (and migration) is phased.

For example, you may decide to develop an initial MVP solution, roll it out to the Australian and New Zealand organization, and then roll it out further to other regions or countries – in bulk or groups of countries.

These are the advantages of phasing by geography:

- Less concurrent effort for training and change management

- Less risk the developed solution does not sit well with any one particular country or region – impeding adoption

These are the disadvantage of phasing by geography:

- Risk of optimizing the solution for one country or region and not carrying out intended process harmonization

> Tip!
> As a general rule of thumb, when a capability is developed, tested, and ready for use, prioritize rolling it out for feedback as soon as possible. Improve the solution based on the feedback received, and then roll it out further to as many geographies as possible in order to ensure the solutions built are *adopted by your entire organization* – so you don't risk different geographies using different systems concurrently.

Having covered Salesforce project delivery methodologies, it's time we talk about change.

Determining your change management strategy

Over the course of your Salesforce project, many things will change for your organization. Most importantly, how your employees work will change. How good you are as an organization at managing these changes has a direct and enormous impact on the success of your Salesforce project. You could design and deliver the most technically sophisticated solution, but if your users don't want, need, or know how to use your solution, then your efforts and investments will not bear fruit.

Introducing change management

Change management encompasses all the efforts, initiatives, activities, and tools you use to maximize the positive outcomes of the changes your organization undergoes.

You need to manage change in your Salesforce project to ensure the intended users of your Salesforce solution adopt and use your Salesforce solution. The level of **adoption** is a key metric for measuring the success of your Salesforce project, which we'll cover further in *Chapter 4, Securing Funding and Engaging with Salesforce and Implementation Partners*.

Adoption, however, is the *combined result of the collective efforts* you put into managing change. Let's look at the reach of the changes that may come with your Salesforce project.

Degrees of change in Salesforce projects

Enhancing business capabilities – through implementing Salesforce – *reaches beyond changing the system component* of a business capability.

Implementing Salesforce may change your business capability components in the following ways:

- **Systems**: Potentially replacing a legacy system with Salesforce will impact your overall system landscape, and likely *how your enterprise systems integrate*.

- **Data**: How data is managed, stored, distributed, consumed, and utilized is affected as part of a Salesforce project. In today's **data-driven enterprises**, data is considered to be one of the greatest assets. You need to consider and map the stakeholders impacted by replacing the old way of working (legacy systems) with Salesforce and design your future data management and governance, and understand the efforts required for building new reporting and analytics capabilities and assets. Engage an **enterprise architect** and a **data architect** to support you in this exercise.

- **Processes**: Your customer-facing and interacting employees' business processes will change. You will – throughout your Salesforce project – focus the bulk of your change management efforts on how this change affects your employees.

- **People**: As systems and processes change and are optimized, the people who previously worked on rudimentary data entry and manual repetitive tasks, will have time freed up to focus on more value-adding activities delivering better business outcomes – all in line with the overall vision of your Salesforce project.

  - There may be a gap between your employee's current competencies and skills compared to those required for the value-adding activities. This will require upskilling and training – beyond learning to use your Salesforce solution. The nature of the upskilling and training, of course, depends on what competency gaps you identify. If you are changing your company's overall sales methodology – for example, from traditional transactional to solution selling – as part of your Salesforce project, you may need to carry out a series of sales training sessions.

  - As part of some projects, you may find some employees that need to learn new skills to work in the new way are not interested in and/or capable of change. In this case, you need to consider where they can potentially fit within your organization in order to still retain the organizational knowledge these employees possess. As a last resort, you may need to consider letting some employees go if they are not interested in what you can offer or if you are unable to find opportunities for them within your organization.

- Work closely with your human resource business partner when carrying out the exercise of identifying changes to different groups of employees to capture changes in job descriptions and required skills.

While each of these elements may currently be the responsibility of different departments within your organization, ensuring they are all addressed is a critical prerequisite for being able to design and deliver solutions your user will adopt. Later on in this chapter, we'll look at how to anchor the collective responsibilities for changing organizational capabilities.

Identifying stakeholders impacted by change

When implementing Salesforce, change will occur at many levels of your organization and the management of those changes will require different types of activities.

In the previous section, you identified the reach of change as a result of implementing Salesforce. Now let's take a look at *who* specifically will be affected by change so you can plan communication and change management activities accordingly:

- **End users**: Their way of working will change as they will work in Salesforce.

- **Managers of your end users**: Salesforce will provide new insights into their employees' work. Managers' buy-in is key for sustaining change and adoption of your Salesforce solution. It's likely the managers will require upskilling and coaching to enable them to support and encourage their employees to adopt your Salesforce solution.

- **Leadership of local sales organizations**: Local leadership will effectively act as local executive sponsors of your Salesforce project. Their buy-in and communication are vital for adoption.

- **Global leadership**: Assuming you will have one **Salesforce org** that is to be used throughout your company, your global leadership will effectively be able to gain insights into marketing, sales, and customer service performance in a whole new way. This will require training and potentially upskilling to gain maximum value for this new level of insights.

- **Application management staff**: The employees who may have managed your legacy CRM system(s) will go through a major change having to learn to manage your Salesforce solution and the users of it.

- **Data, BI, reporting, and analytics staff**: These functions may or may not be part of your IT organization. Regardless of organizational location, their underlying data architecture will change as they – with Salesforce – will have a new source of data as well as a new place to create and present reports and dashboards for reporting stakeholders.

Timing your change management activities by Salesforce project phase

The methods and activities for managing change – your toolbox – are the same regardless of which delivery methodology you choose for your Salesforce project. The *timing* and dosage of your change management initiatives, however, are dependent on which delivery methodology you choose.

Let's explore how the timing of initiatives commonly varies by delivery methodology:

	Delivery methodology		
Key change management initiatives	**Waterfall**	**Hybrid agile**	**Pure agile**
Identify stakeholders impacted by change	Before development starts	Before development starts	At product backlog refinement or sprint planning
Use vision to determine messaging and craft a communication plan	Before development starts	Draft before development starts, adapt continuously	At product backlog refinement
Recruit change agents	Before UAT	Before development starts	In sprint planning
End user involvement	Not until UAT and training	In pre-development phase workshops, in the development phase at show and tells	Continuously at show and tells, and user feedback forums
User training	In roll-out phase	In roll-out phase	At product demos and in application
Monitor and ensure adoption	In roll-out phase	In roll-out phase	Continuously

Table 3.2 – Timing of key change management initiatives by the delivery methodology

Next, let's see what frameworks are available for you to use.

Using a change management framework

There are plenty of change management frameworks for you to choose from. Some of the most commonly used frameworks include the **Prosci ADKAR® Model**, **Drucker School of Organizational Change Leadership**, and the classic **John Kotter's 8-Step Process for Leading Change**.

> **Tip!**
> While detailing the contents of each change management framework is beyond the scope of this book, I encourage you to read about them and be inspired. It is beneficial to follow an overall tried and tested framework but be mindful not to lock yourself into a too rigid structure – incorporate elements you believe to be valuable for your own change management strategy in accordance with your company's culture.

We know your organization will undergo change because that is the goal of your Salesforce project, so we can predict it. What we cannot anticipate is the nature of how the organization – the people affected by the change – will react and respond.

You will spend a great deal of effort throughout your Salesforce project to understand your employees' current processes, and determine future, improved, or new processes to understand what the change will be. We cannot for certain know whether these assumptions about improvements and change will hold true – or even if they are true, whether or not it will be *perceived* as a positive change by the people affected.

> **Tip!**
> Adapt and refine your change management strategy as you progress through your Salesforce project.

Next, we will move on to understanding how to govern and manage your Salesforce program and project.

Envisioning your Salesforce Center of Excellence

Your project taskforce has, until this point, consisted of three key roles (refer to *Chapter 1, Creating a Vision for Your Salesforce Project*), who have done an amazing job getting to this point. Now is the time to envision how to manage your Salesforce program and envision the structure of your **Salesforce Center of Excellence** (**CoE**). Your Salesforce taskforce members will continue to play key roles in your new Salesforce CoE.

Understanding the purpose of a Salesforce Center of Excellence

Your Salesforce CoE will be your management and governance framework to oversee your Salesforce program.

The benefits of establishing your Salesforce CoE are multi-faceted:

- To ensure *timely delivery* of Salesforce projects or features, depending on your chosen delivery methodology

- To ensure your overall *Salesforce solution is compliant* with regulations that your company is subject to

- To enable *faster delivery of innovations and reduce risk*

When executed well, the sum total of these benefits will have a significant financial impact in terms of huge cost savings compared to not having a Salesforce CoE in place.

Drafting the charter for your Salesforce CoE

Create a **charter** for your Salesforce CoE to align your program stakeholders and the Salesforce CoE members.

Your Salesforce CoE charter should contain the following content:

- The vision for your Salesforce program

- The goal and responsibilities of your Salesforce CoE

- Metrics to measure the success of your Salesforce program

Let's look at the responsibilities of your Salesforce CoE.

Determining the structure and responsibilities of your Salesforce CoE

Your Salesforce CoE contains the following elements and groups – each with their own areas of responsibility – collectively working towards the same goal.

Let's look at the tasks behind the responsibilities of the CoE:

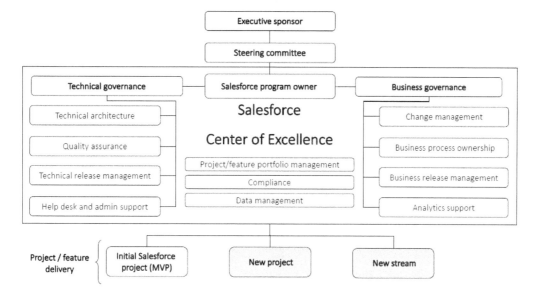

Figure 3.3 – Salesforce Center of Excellence

Executive sponsor: A Senior leadership representative of your company with a vested interest in the success of your Salesforce program. It may be an SVP reporting directly to your company CEO, or sometimes your CEO themself, depending on the size and structure of your organization.

- **Steering committee**: A group responsible for the tactical support of your Salesforce program. It should consist of your executive sponsor, CFO, CIO, *and* senior commercial leadership representatives. You may, at times, wish to invite external parties, such as Salesforce or an implementation partner to present at your steering committee meetings.

- **Salesforce program owner**: Responsible for overseeing and orchestrating the responsibilities of the Salesforce CoE. The program manager drives Salesforce project portfolio management and owns the program roadmap based on input from both the business and technology organizations. The program owner is also responsible for overseeing the delivery of Salesforce projects, though there should be a dedicated Salesforce project manager – or scrum master – for each project or delivery stream.

- **Business governance**: This group is responsible for gathering and consolidating requirements, business process definitions, change management, reporting support, and the business side of release management.

- **Technical governance**: This group is responsible for ensuring your Salesforce org is secure, and its applications are scalable and performant. Furthermore, the group oversees areas of quality assurance, technical release management, as well as help desk and admin support, which includes user management.

The groups and responsibilities in your Salesforce CoE may vary depending on the specific nature of your Salesforce program, industry, and organization.

In the pre-development phase, your Salesforce CoE comes alive and is initiated the moment you secure funding for your initial Salesforce project.

Drafting your Salesforce CoE charter and structure are important prerequisites to being able to create your business case to secure funding for your Salesforce project and program. Once you secure funding, you will want to initiate staffing your Salesforce CoE.

Next, let's look at how to create your **Salesforce roadmap**.

Creating a roadmap for your Salesforce program

The term *program* is used here intentionally as the program may consist of many projects and streams related to Salesforce – but not necessarily included in the scope of your initial Salesforce project, or at least not in the initial release of your Salesforce solution.

Let's see the **Salesforce program roadmap** our scenario company, PME, has put together, illustrated in a classic **transformation map (T-Map)**.

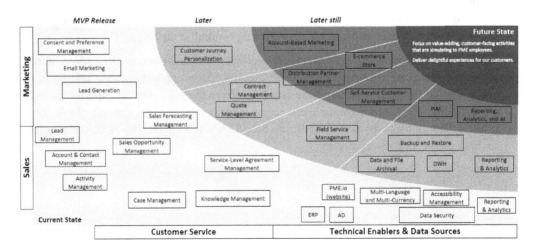

Figure 3.4 – PME's Salesforce program roadmap

PME's Salesforce taskforce put together its road map. They considered project phasing and delivery methodology for their initial release. They decided to keep things simple by creatively defining three buckets for their desired capabilities:

- **MVP Release**: *The MVP release contains capabilities currently supported by PME's existing email marketing and CRM systems, in addition to a few select capabilities that PME believes will bring great value to the business.*

- **Later**: *These capabilities are much desired, yet are dependent on having a number of foundational capabilities in place before starting to build them. PME does not intend to batch the Later bucket capabilities into a single project, they rather intend to distinguish the timing of delivery between capabilities from the MVP Release and the Later still buckets.*

- **Later still**: *These capabilities are also desired, yet still dependent on capabilities delivered in the MVP Release and the Later buckets.*

PME has effectively phased its Salesforce implementation by scope, already reducing the risk of its implementation significantly. It is also keen to phase its project by geography.

Iterating in the pre-development phase

Determining how to deliver your Salesforce project is an important undertaking. Determining your Salesforce program roadmap and the scope for an initial Salesforce project is an iterative process. You will return to your roadmap and make modifications as you become aware of dependencies and more familiar with the cost to acquire licenses and implement features. Iteration and modifications are great – you are further shaping your Salesforce to fit your organization. Capture any changes and decisions in your RAID log.

Summary

In this chapter, you have made many considerations for how best to deliver your Salesforce project. You have considered what delivery methodology is most appropriate for your particular project and organization, and you have considered how you might phase your implementation by scope and geography to reduce the risk and increase the speed of delivery. Then, you were introduced to change management, the degrees of change management. You also learned how to identify the many stakeholders affected by change as a result of your Salesforce project, and how you should *time* your change management activities depending on your chosen delivery methodology. Finally, you envisioned your Salesforce CoE, your management and governance body to oversee your Salesforce program, and you learned how to create a Salesforce program roadmap.

Up until now, you have created a vision for your Salesforce project – the *why* – as well as the nature of it – the *what* – of your project, and you have just determined *how* to deliver it.

You are now ready to move on to the last chapter in the pre-development phase of your Salesforce implementation: *Chapter 4, Securing Funding and Engaging with Salesforce and Implementation Partners.* It is time to determine *how much*.

4

Securing Funding and Engaging with Salesforce and Implementation Partners

In this chapter, we enter the financial and legal dimensions of Salesforce projects. First, we will cover how to create and present a **business case** to secure funding for your project. Next, we explore engaging with Salesforce to get a quote for your license costs, and we'll go through the intricacies of communicating with implementation partners, and what to look for when selecting one.

You will be introduced to methods of **financial forecasting**, including how your Salesforce project will increase your company revenue and diving into what costs you need to consider – and forecast – as part of your Salesforce implementation. You will learn what roles you are likely to encounter when engaging with Salesforce and how to best request a quote from Salesforce. Finally, you will learn about the value of working with an implementation partner, what types are available, the **selection process**, and how to **contract** with one.

After securing funding for your Salesforce project, acquiring Salesforce licenses from Salesforce, and contracting with an implementation partner, you'll finally be ready to start the development phase of your Salesforce implementation.

This chapter will cover the following main topics:

- Creating a business case to secure funding
- Engaging with Salesforce and implementation partners
- Iterating in the pre-development phase

Let's go!

Creating a business case to secure funding

In the previous chapter, you determined how you are going to deliver your Salesforce project – the delivery methodology, communication, and the change management strategy – and envisioned your Salesforce CoE to govern your Salesforce implementation. It is now time to secure funding for your project and begin to engage with Salesforce and implementation partners, as shown in the following diagram:

Figure 4.1 – Activities in the pre-development phase

Before being able to create the financial forecast, let's get specific about what business outcomes you are expecting from your Salesforce project.

Determining KPIs and targets for your required business capabilities

If you want to be able to measure whether your Salesforce project is a success when it is rolled out, you need to establish how you will measure performance and what targets define success. Your Salesforce taskforce should not be setting the business targets; you should be engaging with your business stakeholders to determine what **key performance indicators** (**KPIs**) make sense to track and what **KPI targets** to aim for.

This may seem like a big undertaking upfront but consider the opposite: if no KPIs or goals are set, project success measurement will rely on subjective qualitative input from various user groups rather than pre-determined and well-founded considerations. Furthermore, you will be thrilled to have done this exercise when you put together your business case and engage with Salesforce and implementation partners.

In the following table, you will see some examples of how **Packt Manufacturing Equipment (PME)** is determining KPIs and targets for their business capabilities:

Domain	Capability	KPI	KPI Target
Marketing	Lead generation	Number of marketing-qualified leads generated	10% increase
Sales	Lead management	Average lead processing time	35% decrease
		An increased lead conversion ratio	5% points increase For example, from a 25% to 30% conversion ratio
	Opportunity management	Number of new opportunities created by sales	3% increase
		Average duration from opportunity creation to closed-won	20% decrease
		Average closed-won opportunity ratio	2% points increase
Customer service	Case management	**Average handle time (AHT)**	20% decrease
		Case resolution time	15% decrease
		Net promoter score (NPS)	10% increase

Table 4.1 – Examples of KPIs and targets for business capabilities in PME's Salesforce project

> **Tip**
> If your company is currently not tracking these KPIs, your first task – as a member of the Salesforce taskforce – is to assist your business stakeholders in establishing a valid baseline before determining the targets for your Salesforce project.

Having determined your KPIs and their targets, let's move on to your business case.

Creating the business case for your Salesforce project

Implementing Salesforce as a whole needs to make financial sense by putting your company in a better financial position compared to the status quo, as well as compared to implementing an alternative CRM solution. Recall how we distinguished between programs and projects in the previous chapter.

As such, your initial Salesforce project, which will consist of a smaller scope than what you envision your total solution will be, should also deliver a better financial position independently.

In this section, we'll go through the exercise of creating the business case for your initial project. While the elements will differ depending on scope, the model is fundamentally the same.

Regardless of the business problem you are addressing, all enterprise CRM projects requesting funding should be able to articulate their value by at least one of the following factors:

- Increasing revenue
- Reducing cost
- Having the net sum of impact on revenue and cost be greater than zero

> **Tip**
> Discuss with your CFO or financial business partner what criteria your company's **investment board** focuses on when evaluating funding requests.

In addition to your financial forecast, there are other elements to include in your business case. We'll return to that later in this section. For now, let's see how you can estimate *increased revenue* as a result of implementing Salesforce.

Estimating the revenue impact of your Salesforce project

Implementing Salesforce can impact multiple business domains:

- **Marketing**: More and better leads and traffic to e-commerce stores.
- **Sales**: Productivity gains from 360-degree customer view, automation, and integration means sales reps can close more deals faster.
- **Service**: While historically being considered a cost center in the enterprise, companies are increasingly realizing the essential role the customer service organization (customer success) plays in customer satisfaction and the length of the customer's lifetime.
  - Some companies are also commercializing their service offerings. The methods range from simply charging for access to premium customer service (for example, some low-cost airlines charge for customer service), to selling service entitlements, and some companies are completely integrating (bundling) services into their product offerings.
  - Service companies have done this for years, but other industries, especially companies in the manufacturing industry, have enormous potential to add another revenue stream or increase the value to their customers by including (or highlighting) service in their total offered solution.

- **Innovation**: Implementing Salesforce will benefit your company in more ways than direct revenue or cost impact. Implementing a cloud-based, state-of-the-art platform will allow your company to innovate faster, implement new processes, react and pivot to market changes, quickly integrate newly acquired companies, and divest business units. Moreover, a well-architected Salesforce solution – as you will strive to have – will also reduce operational risk to business continuity as your organization will be less prone to system downtime (versus legacy systems). While it's harder to pin down the exact value of a company's innovation capability and reduced operations risks, it is a valuable benefit that should be highlighted in your business case.

> **Important note**
>
> With regard to estimating the value of improved innovation capability, you have two options: either attempt to estimate the value of your increased innovation capability's impact on revenue and cost – adding 1-2% to revenue impact and reducing operating costs similarly – or you can simply highlight the *innovation benefits* in your business case. I recommend the latter approach as it keeps items you can estimate *separate* from the more abstract items.

Next, we'll look at costs.

Estimating the cost impact of your Salesforce project

As you have learned in *Chapter 2, Defining the Nature of Your Salesforce Project*, implementing Salesforce has wide-reaching business impact – which is the goal. On the other hand, your Salesforce project will also have an impact on many other costs – with both positive and negative financial impacts. For your Salesforce project, you need to consider the following costs:

- **New license costs**: The cost of procuring new Salesforce licenses for your users.

- **Saved license costs**: Cost savings from retiring legacy systems.

- **Implementation costs**: The costs to deliver your Salesforce project include the costs to *develop*, *deploy*, and *roll it out*, as well as the *critical communication and change management* activities you will do throughout your Salesforce project. These costs are tricky to estimate at this point since you may not yet have engaged with any implementation partners. Once you do, come back to this element, and update accordingly. For now, you should connect with an enterprise architect or IT business partner at your company who will have some insights into comparable cloud projects.

- **Operating costs**: The impact your Salesforce project will have on staff costs.

  - **Business operations costs**: These include savings from productivity gains of your business front-office staff (sales and customer service functions) by having a 360-degree view of their customers, automations, and integrations. Time saved by employees can either be converted to cost savings or directed to more value-adding activities.

- **Salesforce CoE costs**: New staff to govern, maintain, and run your Salesforce solution (non-project costs). These include internal admins and business analysts who carry out minor bug fixing, implement small features requests, and user management – *keeping the lights on*. We'll cover this in more detail in the next section.

- **Legacy system operating costs**: Staff who won't be maintaining legacy systems being retired. If possible, you should retrain and upskill these people to be part of your Salesforce CoE as they are intimately familiar with your organization and possess many transferable skills, which – combined with Salesforce knowledge – will be a valuable resource for your new Salesforce CoE. As Salesforce may be an entirely new platform and cloud computing may be new to these people, be mindful and diligent in training – we'll return to this topic in *Chapter 12*, *Evolving Your Salesforce Org and DevOps Capabilities*.

Let's look at your options for implementation partner models.

Choosing your implementation partner model

Conway's law says organizations, if not provided external input, will more or less lift and shift existing processes (and data) into new systems. And what a waste of a perfectly good opportunity if your company invests in the shiny new, state-of-the-art, market-leading CRM platform that Salesforce is without improving its processes.

This is where Salesforce implementation partners come into the picture.

Let's look at the different partner models:

- **100% insourcing**: Your company does not work with an implementation partner at all. For Salesforce projects, this is not a recommended approach for initial greenfield implementations. After initial implementation is deployed, bug fixes and feature enhancements to existing clouds may be carried out by an internal team, supported by clearly defined development guidelines and governance.

- **100% outsourcing**: Your company outsources everything except some high-level governance and providing requirements. The implementation partner drives development, deployment, training, user management, and the help desk. There are valid reasons for choosing 100% outsourcing – such as if your company's IT resources are limited, but you want to develop great solutions and provide timely support for your users.

- **Happy medium**: In this model, you handle some parts of Salesforce yourself – most often elements such as user management, reporting support, bug request management, new feature request management, training, and the help desk. There are endless variations to this model, and it is also the most prevalent. What you need to make clear is what elements your company will be better at than an implementation partner and for what reason. Typically, this comes down to the elements that involve communicating with your end users.

The partner model you choose is not set in stone. As you progress through the phases of your Salesforce implementation, learn what works well and what doesn't, and optimize your setup accordingly.

> **Important note**
>
> The cost for implementation partners falls under your company's **capital expenditures (CAPEX)**, whereas your daily internal operating expenses fall under **operating expenditures (OPEX)**. Discuss with your CFO or other finance business partner whether your company has a preference or considerations in relation to choosing a partner model for your Salesforce implementation.

Let's move on to discuss the resources required in your Salesforce CoE.

Determining your Salesforce CoE target operating model

You need to include the cost in terms of time spent by the members of your Salesforce CoE in delivering your Salesforce project. To determine that, we first need to look at what roles need to be part of your Salesforce CoE.

To help you determine what roles you need to purely *maintain* your implementation – after initial implementation and roll-out – review the Salesforce help article *BEST PRACTICES: Achieve Outstanding CRM Administration*, at `https://help.salesforce.com/s/articleView?id=000327119&type=1`.

The model in the article is based on up to 750 users. For enterprise organizations planning to invest in and implement Salesforce, it is critical to know what key factors play into determining the required resources to run their Salesforce platform to be able to estimate the operating cost to maintain their solution.

So, let's dive into these considerations and factors.

In all Salesforce orgs, you will have at least one internal user who will have administrator rights for configuring your Salesforce org. How many and when depends on a number of factors and decisions around your operating model – and your chosen partner model.

The following list includes some of these factors:

- Whether your Salesforce org is live in production or not
- The breadth of Salesforce clouds and add-on products used
- The degree of customization of your Salesforce org
- The maintainability of the solutions* in your Salesforce org

> *Tip
>
> Maintainability is a factor you *can influence* by having the right governance mechanisms in place for your Salesforce project and program. We'll cover this in more detail in *Chapter 7, Building and Testing Your Initial Release.*

- The number of internal Salesforce users

- The number of lines of business in your Salesforce org

- The number of sales regions or countries

- The number of customer-facing applications such as Experience Cloud sites or commerce storefronts

Some of the factors are purely based on facts, for example, the number of users.

Your Salesforce CoE members are responsible for areas other than user management – recall the Salesforce CoE structure diagram from the previous chapter. Not all members will be 100% allocated to the Salesforce CoE. Therefore, you should use an **activity-based costing model** to calculate the costs to include in your business case for your Salesforce program.

Let's see what PME's Salesforce CoE member roles and time allocation look like:

CoE Role	# of People	Time Allocation To Salesforce CoE	Annual Salary	CoE ABC Cost
Executive Sponsor	1	20%	$ 300,000	$ 60,000
Program Owner and MVP Project Manager	1	100%	$ 200,000	$ 200,000
Product Owner	1	100%	$ 150,000	$ 150,000
Admin	3	100%	$ 100,000	$ 300,000
Enterprise Architect	1	10%	$ 200,000	$ 20,000
Data Architect	1	20%	$ 150,000	$ 30,000
Domain Subject Matter Experts (SMEs)	3	25%	$ 150,000	$ 112,500
Change Agents (From Local Sales Organizations)	8	20%	$ 150,000	$ 240,000

Annual Salesforce CoE Cost   **$ 1,112,500**

Figure 4.2 – PME Salesforce CoE roles, time allocations, and costs

PME has 2,000 full-time employees, of which half will work in Salesforce and require licenses and user management. PME's enterprise architect will advise how Salesforce fits within PME's business strategy, and the data architect will provide expert input on how Salesforce can support PME's data strategy and execution of it. The domain subject matter expert and change agents will participate throughout the project to help increase the chances that the solution meets business requirements and that it will be well-received by PME business users.

It may seem like a lot of people and cost, however, the value of a strong, aligned Salesforce CoE will pay dividends down the road.

Having completed the initial exercises of determining your KPIs and their targets, as well as the estimated cost of the Salesforce CoE, it's time to get out the calculator (or spreadsheet) and crunch some numbers.

Forecasting the financial impact of your Salesforce project

To forecast how your Salesforce project will impact your company financially, you need to start by consolidating your KPIs, their targets, and your presumptions and assumptions.

Let's see the presumptions and assumptions PME have for their Salesforce project:

	Presumptions and assumptions
Annual Revenue	Product of Marketing Impact and Sales Impact – fully phased 18 months after last roll-out.
Sales Impact on Revenue	
Annual Number of New Opportunities	Derived from PME company information and results of the KPIs and KPI target definition exercise. Increased number of new opportunities as a result of better insights, customer 360-degree overview, more efficient opportunity management.
Average Closed Won Opportunity Ratio	
Number of Closed-Won Opportunities	
Average Closed-Won Opportunity Amount	
Closed-Won Opportunity Value = Revenue Impact From Sales	
Marketing Impact on Revenue	
Annual Number of Marketing Qualified Leads Generated	Derived from PME company information and results of the KPI and KPI target definition exercise. Increased number of new opportunities as a result of more Marketing Qualified Leads (MQLs) generated due to better marketing automation. In order to measure the isolated effect of better markting automation, we are not including the drivers for Sales Impact on Revenue from above.
Lead Conversion Ratio	
Number of Converted Leads = Number of New Opportunities	
Average Closed-Won Opportunities Ratio	
Number of Closed-Won Opportunities	
Average Closed Won Opportunities Amount	
Closed-Won Opportunity Value = Revenue Impact From Marketing	
Gross Margin %	No impact on Gross Margin %.
New License Costs - Salesforce	Costs incurred from acquisition of licenses just in time for go-live for each PME local sales organization.
Salesforce Enterprise Edition License Costs	1,000 PME Sales and Service staff with full CRM license at list price for Enterprice Edition. https://www.salesforce.com/editions-pricing/sales-and-service-cloud/
Salesforce Marketing Cloud Account Engagement (MCAE) License Costs	MCAE Plus edition license cost – list price. 5 extra Contact Blocks to accommodate 120,000 contacts. https://www.salesforce.com/editions-pricing/marketing-cloud/account-engagement/
Legacy System License Costs	From go-live of Salesforce plus 3 months. Assumption: Cost model of legacy system is per-user.
Operating Costs	Product of below impacts.
Business Operations Cost: Sales and Customer Service	Derived from PME company information and results of the KPI and KPI target definition exercise.
Salesforce CoE Costs	Costs incurred as per PME Salesforce CoE roles and time allocations by project phase.
Legacy System Operating Costs	Costs incurred simultaneously with saved "Legacy system license costs".
Investment (CAPEX): Implementation Costs	One-off cost based on PME's enterprise architect's estimate for implementation and local deployment roll-outs.
Operating Profit: Gross Margin less OPEX and CAPEX	Net impact of revenue and cost impact.

Figure 4.3 – Presumptions and assumptions for PME's Salesforce project

Next, you need to translate all your presumptions and assumptions into a financial model to compare the impact of your Salesforce project to your company's baseline performance.

> **Important note**
>
> Although your Salesforce program roadmap contains more capabilities to be added to your Salesforce solution over time, *you should only include* the financial impact of the scope included in the project for which you are preparing this business case.

In the following figure, you can see PME's financial model for the impact of their Salesforce project:

PME's Salesforce Project Impact vs. Baseline	Baseline	% change vs. Baseline	Future		Absolute change vs. Baseline
Annual Revenue	$ 300,398,180	12.7%	$ 338,688,553	$	38,290,373
Sales Impact on Revenue					
Annual Number of New Opportunities	20,000	3.0%	20,600		600
Average Closed Won Opportunity Ratio	23%	2%-points	25%		N/A. Percentage
Number of Closed-Won Opportunities	4,600	12.0%	5,150		550
Average Closed-Won Opportunity Amount	$ 63,500	None	$ 63,500	$	-
Closed-Won Opportunity Value = Revenue Impact From Sales	$ 292,100,000	12.0%	$ 327,025,000	$	34,925,000
Marketing Impact on Revenue					
Annual Number of Marketing Qualified Leads Generated	2,200	10.0%	2,420		220
Lead Conversion Ratio	18%	5%-points	23%		N/A. Percentage
Number of Converted Leads = Number of New Opportunities	396		557		161
Average Closed-Won Opportunities Ratio	33%	None	33%		N/A.
Number of Closed-Won Opportunities	131		184		53
Average Closed Won Opportunities Amount	$ 63,500	None	$ 63,500	$	-
Closed-Won Opportunity Value = Revenue Impact From Marketing	$ 8,298,180	40.6%	$ 11,663,553	$	3,365,373
Gross Margin %	45.0%	None	45%		N/A.
Gross Margin Impact	$ 135,179,181	12.7%	$ 152,409,849	$	17,230,668
Operating Costs (OPEX)	$ 100,500,000	-0.6%	$ 99,871,500	$	(628,500)
New License Costs - Salesforce	$ -	N/A new	$ 1,719,000	$	1,719,000
Salesforce Enterprise Edition License Costs			$ 1,680,000		
				$	1,680,000
Salesforce Marketing Cloud Account Engagement (MCAE) License Costs					
			$ 39,000	$	39,000
Legacy System License Costs	$ 1,500,000		$ -	$	(1,500,000)
Operating Costs	$ 99,000,000		$ 98,152,500	$	(847,500)
Business Operations Cost: Sales and Customer Service	$ 48,000,000	-2.0%	$ 47,040,000	$	(960,000)
Salesforce CoE Costs	$ -	N/A new	$ 1,112,500	$	1,112,500
Legacy System Operating Costs	$ 1,000,000	-100%	$ -	$	(1,000,000)
Other Operating Costs	$ 50,000,000		$ 50,000,000	$	-
Investment (CAPEX): Implementation Costs	$ -		$ 7,440,000		N/A one-off
Operating Profit: Gross Margin Less OPEX and CAPEX	$ 34,679,181	30.0%	$ 45,098,349	$	10,419,168
Operating Profit %	11.5%	15.3%	13.3%		N/A.

Figure 4.4 – PME's Salesforce project impact versus baseline

Now, as we mentioned in *Figure 4.3*, some of the impacts are phased over a period of months – for example, productivity gains don't happen overnight, and neither are legacy system license costs saved until you at least can deactivate users when they start using your new Salesforce solution.

> **Tip**
> To best manage the expectations of your executive sponsor and your steering committee members, be realistic and conservative: phase revenue impact and productivity gains over time. Incur costs upfront.

Let's look at a **profit and loss** (**P&L**) impact forecast example from the PME Salesforce taskforce – taking impact phasing into account:

	Profit & Loss Impact '000						
	Year 1	Year 2	Year 3	Year 4	Year 5	3-year Total	5-year Total
Annual Revenue Impact	$ -	$ 4,964	$ 28,541	$ 38,290	$ 38,290	$ 33,504	$ 110,085
Gross Margin %		45%	45%	45%	45%	45%	45%
Gross Margin Impact	$ -	$ 2,234	$ 12,843	$ 17,231	$ 17,231	$ 15,077	$ 49,538
Operating Costs (OPEX) Impact	$ 1,171	$ 407	$ (205)	$ (389)	$ (389)	$ 1,373	$ 596
New license Costs - Salesforce	$ 69	$ 1,367	$ 1,719	$ 1,719	$ 1,719	$ 3,155	$ 6,593
Salesforce Enterprise Edition License Costs	$ 39	$ 1,330	$ 1,680	$ 1,680	$ 1,680	$ 3,049	$ 6,409
Salesforce MCAE* License Costs	$ 30	$ 37	$ 39	$ 39	$ 39	$ 106	$ 184
Legacy System License Costs Impact	$ (6)	$ (1,188)	$ (1,500)	$ (1,500)	$ (1,500)	$ (2,694)	$ (5,694)
Operating Costs Impact	$ 1,108	$ 228	$ (424)	$ (608)	$ (608)	$ 912	$ (303)
Business Operations Costs: Sales And Customer Service Impact	$ -	$ (93)	$ (537)	$ (720)	$ (720)	$ (630)	$ (2,070)
Salesforce CoE Costs	$ 1,113	$ 1,113	$ 1,113	$ 1,113	$ 1,113	$ 3,338	$ 5,563
Legacy System Operating Costs Impact	$ (4)	$ (792)	$ (1,000)	$ (1,000)	$ (1,000)	$ (1,796)	$ (3,796)
Other Operating Costs Impact	$ -	$ -	$ -	$ -	$ -	$ -	$ -
Investment (CAPEX): Implementation Costs	$ 5,252	$ 2,188	$ -	$ -	$ -	$ 7,440	$ 7,440
Impact On Operating Profit: Gross Margin Less OPEX And CAPEX	$ (6,423)	$ (362)	$ 13,048	$ 17,619	$ 17,619	$ 6,264	$ 41,502
Cumulative Impact On Operating Profit	$ (6,423)	$ (6,785)	$ 6,264	$ 23,883	$ 41,502	$ 6,264	$ 41,502
Payback Time	32 Months						
Return On Investment (ROI)						84%	558%

*Marketing Cloud Account Engagement

Figure 4.5 – P&L impact of PME's Salesforce project

In addition to calculating the P&L impact of your Salesforce project, you need to determine two key financial investment metrics:

- **Payback time**: The *duration* before the negative financial impact of your investment – Salesforce licenses, implementation and roll-out costs, and your new Salesforce CoE – is weighed up (paid back) by the positive financial impact on your operating profit from the expected superior business outcomes: increased revenue and reduced operating costs.

- **Return on investment (ROI)**: A simple *ratio* of your one-off investment versus the impact on your operating profit. The one-off investments in the case of PME are implementation and roll-out costs.

Since organizations differ in risk appetite, what an acceptable payback time is, and how long it is willing to wait to reap the financial benefits of its investment in Salesforce, it is advisable to include both 3- and 5-year financial projections.

You have done all the groundwork and created a great initial business case. Now it's time to present it to your company's leadership to secure funding for your Salesforce project.

Presenting your Salesforce project business case

If you have been following each chapter in the pre-development phase section of the book, creating your **business case presentation** will be a breeze.

Invite these key stakeholders for your meeting: the executive sponsor, **Chief Executive Officer** (**CEO**), **Chief Information Officer** (**CIO**), **Chief Financial Officer** (**CFO**), representatives from commercial leadership, the enterprise architect, and all your fellow Salesforce taskforce members.

Include the following sections in your Salesforce project business case presentation:

- **Executive summary**: Include the key financial metrics of your project as well as the benefits of improving the innovation capabilities of your company.

- **Vision**: Include the vision you created in *Chapter 1, Creating a Vision for Your Salesforce Project*.

- **Nature of the project**: Include the business domain capabilities and the hypothetical technical scope you defined in *Chapter 2, Defining the Nature of Your Salesforce Project*. Highlight how **business users** will stand to benefit from your project and how it will improve the **customer experience**.

- **Project governance**: Show the governance and management framework you envision your Salesforce CoE will provide referring to your work on this in *Chapter 3, Determining How to Deliver Your Salesforce Project*.

- **Delivery methodology and phasing approach**: Explain how you plan to deliver your Salesforce project as defined in *Chapter 3, Determining How to Deliver Your Salesforce Project*.

- **Indicative project timeline**: Depending on your chosen delivery methodology and phasing approach – in combination with your estimate of implementation costs – highlight the following milestones of your Salesforce project: project start, development start, UAT, first go-live, and the target month or quarter when all intended users are live on your solution.

- **Change management strategy**: As you determined in *Chapter 3, Determining How to Deliver Your Salesforce Project*.

- **Indirect benefits**: Highlight the benefits of being able to innovate, adapt to change, and lower risk to business continuity.

- **Implementation cost**: Refer to your work on financial forecasting earlier in this chapter.

- **Project investment**:

  - 3-year and 5-year financial impact

  - Project payback time (month)

  - Project ROI

- **Program roadmap**: Looking beyond your Salesforce MVP solution, include the roadmap you created in *Chapter 3, Determining How to Deliver Your Salesforce Project*.

You have now presented your business case to management – well done!

If it was the first meeting, and you have yet to engage with Salesforce and implementation partners, you will have based your license cost and implementation costs on assumptions and previous experience.

If you didn't secure funding in the first meeting, ask for and note down the key objections and feedback. This will let you know which areas you need to work on to make the business case stronger – you will need to consider what to change and how it changes other components and costs. Potentially modify the scope, delivery methodology, phasing, KPIs and targets, and, therefore, the financial forecast.

You may find that the funding for your Salesforce project is not approved because senior management believes your business case is reliant on assumptions of costs but that you are approved to begin engaging with Salesforce and implementation partners to solidify your assumptions. You will gain insights and assistance for this in the next section, where we explore how you can effectively engage with Salesforce and implementation partners.

Now, it is time to talk with Salesforce and implementation partners.

Engaging with Salesforce and implementation partners

Once a budget for your Salesforce project has been pre-approved, you can begin to engage with both Salesforce and implementation partners simultaneously. In this section, we'll cover the basics of how you can do so effectively.

Engaging with Salesforce

To validate your business case assumptions for what your Salesforce license costs will be, you need to get a quote from Salesforce.

Let's see who you will engage with to get your quote.

Parties involved from Salesforce

When you visit www.salesforce.com and request to watch a product demo or download a whitepaper and fill out your information, you will be contacted within 24 hours – or in most cases within the same business day. A **business development representative** will call you to find out whether you want to speak to an account executive.

You will, as you progress your dialogue with Salesforce, be introduced to a number of different roles:

- **Account Executive (AE):** Depending on the size of your company and what clouds/products you are looking to request a quote for, you will be assigned a **core AE** and one or more cloud-/product-specific AEs, for example, a **Marketing Cloud AE**

- **Solution Engineer (SE):** Works with you to understand your business requirements and translate them into Salesforce products that may meet them

Next, let's see how you get a quote from Salesforce.

Requesting a quote from Salesforce

The best way to request a quote from Salesforce is by recapping who your users are and what they will do in Salesforce. A standard method is by creating an **actor diagram** or table.

Here is an example of an actor diagram for the first phase of PME's Salesforce project:

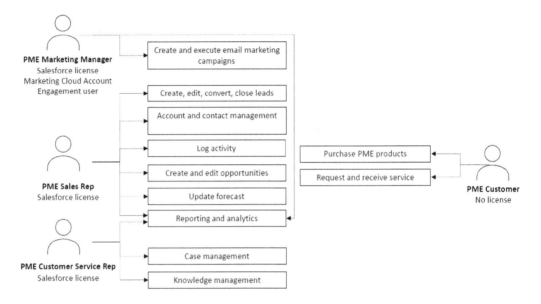

Figure 4.6 – PME actor diagram

A few key additional elements to include in your request are as follows:

- Include the vision you created in *Chapter 1, Creating a Vision for Your Salesforce Project*.

- Include the business domain capabilities and the hypothetical technical scope you defined in *Chapter 2, Defining the Nature of Your Salesforce Project*.

- If you have completed the exercise in the *Org editions* section in *Chapter 2, Defining the Nature of Your Salesforce Project*, include the edition you think you will need. If you haven't, I recommend you go back and read that section.

- Consider when you may actually start the development/configuration of the feature. State when you will need how many of each license – include the roadmap you created in *Chapter 3, Determining How to Deliver Your Salesforce Project*.

Great, you have prepared your request for a quote and sent it to Salesforce. Let's see how Salesforce will respond.

Evaluating your quote from Salesforce

When you speak with an account executive and a solution engineer from Salesforce, they will ask several questions about topics you haven't even considered. They are using these probing questions to understand what products may be relevant for your company.

Be prepared to be amazed and inspired. You may find yourself a bit overwhelmed with all the options and possibilities. Keep a healthy balance of inspiration and trust in the work you have done until this point in terms of delivery methodology, project phasing, and roadmap creation.

You may find yourself updating your various artifacts to include some new capabilities and new Salesforce products.

> **Tip**
> As you engage with implementation partners and are provided with their proposed solutions and required licenses, you may find yourself needing to modify the needed licenses. To settle on the required licenses, have alignment calls with Salesforce and your chosen implementation partner.

Next, let's talk with implementation partners.

Engaging with implementation partners

In this section, we'll cover types of implementation partners, the process of selecting one, how to communicate with them, what to look for when selecting one, and what to pay attention to when contracting.

Types of implementation partners

As of August 22, 2022, there are 2,141 registered consultancies on Salesforce AppExchange (`https://appexchange.salesforce.com/consulting`) offering implementation services. They range from one-man shows to global consulting firms.

Let's understand how the types of implementation partners differ:

- **Global consulting firms**: These are major consulting, IT strategy, and implementation companies. By definition, they cover all types of IT services and domains. They are partners with all the largest enterprise software companies (Salesforce, Microsoft, SAP, Google, and AWS). They also provide custom software development services in all programming languages and have major **offshore** hubs for application and system development and managed services. They typically leverage their scale and offshore hubs for competitive advantage.

- **Specialty firms**: These types of implementation partners focus on Salesforce implementations. They range from small boutique firms to full-fledged consultancies offering strategy and change management advisory services in addition to Salesforce project delivery. Most specialty firms focus on specific clouds and/or industry segments. The specialty firms stand out as niche cloud and/or industry experts.

- **Contractors**: These are individuals who prefer not to have a permanent contract. They typically contract full-time, exclusively, for one company for 3-6 months with the possibility of an extension at the end of the contract period.

- **Freelancers**: They are similar to contractors, but instead of contracting exclusively with one company, they may work with multiple clients at a time.

- **Salesforce Professional Services**: The services arm of Salesforce. Similar to specialty firms as they solely focus on Salesforce implementation services.

Let's see how you might go about finding the implementation partner for your Salesforce project.

The process of selecting an implementation partner

There is a series of steps involved in the process of selecting your implementation partners – each serving their own purpose – as illustrated in the following diagram, where each step will be expanded upon later in this chapter:

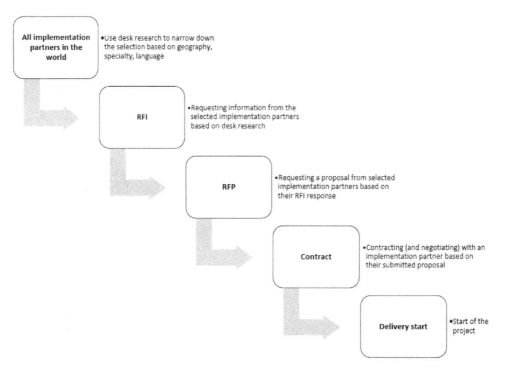

Figure 4.7 – High-level example of a Salesforce implementation partner selection process

Figure 4.7 describes the funnel of finding a suitable Salesforce implementation partner for your Salesforce project. The funnel initially contains more than 2,000 potential implementation partners, which is then reduced in each step until you have contracted with your chosen one and the delivery project can commence.

> **Important note**
>
> Connect with your procurement department to align on the selection process. Procurement processes vary, and detailed procurement processes are not in the purview of this book. It is recommended you engage with your procurement team to understand your company's guidelines and have a category manager/purchasing manager partner with you and go through the process of selecting and contracting with your chosen implementation partner.

Next, let's explore the common practices for engaging with potential implementation partners.

Requesting information from implementation partners

When you send a **request for information** (**RFI**), you are merely requesting information about potential implementation partners. Use an RFI to *pre-qualify* potential implementation partners without providing comprehensive and sensitive information about your company in the request. However, to make the implementation partner understand the seriousness and legitimacy of your request – to increase the likelihood of a proper response – you should include at least the following elements in your RFI:

- Basic information about your company
- The reason for your request
- The requested due date of response
- The information you are requesting – keep this as short as possible

Why share this information at this stage? This is for the implementation partners to be able to internally qualify and assign the request to the appropriate team or person and respond.

Once you have received and compared the responses, it's time to invite some of them to participate in the next step of the process.

Requesting a proposal from implementation partners

Having pre-qualified some of the implementation partners who responded to your RFI, let's continue the process. Next up is sending your **request for proposal** (**RFP**).

> **Tip**
> Given the nature of many RFP responses having 50-100 pages or slides, it is recommended to invite *about 4-7* potential implementation partners, depending on your confidence that they will want to participate in the RFP. For large implementation projects, include both specialty and global consulting firms.

Instead of talking about it, let's look at what an RFP should contain at a minimum for an implementation partner to be able to provide you with a proposal your company can effectively evaluate:

- **Your company background, context, and challenges**: Share a condensed version of the analysis you carried out in *Chapter 1, Creating a Vision for Your Salesforce Project.*

- **Nature of the project**: Include the business domain capabilities and the hypothetical technical scope you defined in *Chapter 2, Defining the Nature of Your Salesforce Project.* Highlight what **business users** need to benefit from your project and how ultimately, it will improve the **customer experience**.

- **Enterprise architecture guidelines**: *Consult the enterprise architect at your company to provide this.*

- **Concurrent programs/projects**: *Consult the enterprise architect at your company to provide this.*

- **Hypothetical delivery methodology**: Include information about the delivery methodology you plan to use, how you understand it, and why you have chosen it. Be open to suggestions and arguments from implementation partners – many implementation partners have a defined model for delivering projects. Refer to *Chapter 3, Determining How to Deliver Your Salesforce Project.*

  - Be sure to include change management and communication services to be part of the proposal response.

- **Key assumptions**: The key assumptions you have made up until this point – *refer to your RAID log.*

- **Implementation partner background**: You should request information about previous projects delivered by the implementation partner in your industry and within the technical scope of your Salesforce project.

- **Proposal response format**: If you have no preference, state that. There are pros and cons to this approach. You risk being overwhelmed by content if you do not specify a format; on the other hand, you get a feel for how the participating implementation partners work and present information.

  - At a minimum, add a description of *what* you are requesting the invited partner's proposal to include in their response, which may comprise the following:

    - Proposed solution architecture

    - Required licenses

- Implementation estimate

- Relevant client reference cases – industry or technical scope

- Team structure

- Project governance

- Project price or rates for proposed roles

- **Health, safety, and environment (HSE) information**: If you and your company subscribe to the notion that people and teams perform their best work when – at a minimum – basic HSE measures are in place, you should require information to support the claim. The request may have many forms depending on your industry, ranging from simple descriptions of how the implementation partner operates and provides adequate working conditions for their employees, to objective and externally accredited certifications, such as from the **International Organization for Standardization (ISO)**.

- **Legal**: Consult your company's legal counsel to determine whether you might require the implementation partners to sign a **non-disclosure agreement (NDA)** prior to receiving your RFP document.

- **RFP process and timeline**: Share the following information:

  - The deadline for submitting a response

  - The process for asking questions about the RFP document and project

  - The process after submission of the response

  - The target contracting date

  - The target start date for the commencement of the project

> Tip
>
> Be as open as you can about the business challenges your company aims to address in your Salesforce project. The more accurate and concise you are, the better the RFP responses will be.

Great, you have assembled your RFP document and sent it to pre-qualified implementation partners. Next, let's look at what to emphasize when evaluating the received proposals.

What to look for in partners

Choosing an implementation partner for your Salesforce project can be overwhelming, and not only because of the sheer volume of information you need to consume and try to comprehend. Common procurement practices state you need to define a number of evaluation criteria for the RFP responses. These often include price, relevant technical expertise, and company risk.

In addition to the usual suspects, I recommend you pay attention to the following characteristics of the implementation partners you have invited to participate in the RFP process:

- Which implementation partner(s) *challenge you*, offer a different or new perspective on your business challenges, and suggest alternative technical solutions or different project phasing approaches?

- *Think end-to-end*, even if you did not explicitly ask them to. For example, if an implementation partner asks several questions regarding reporting strategy or billing, invoicing, and payment processes – just for them to understand the end goal – that is a great sign that they consider your end-to-end business processes.

- Boldly *highlight risks* and propose mitigation actions. Experienced professionals know what risks are common for your type of project and will know how to advise you to proceed.

Let's see what insights PME gained from the responses to their RFP.

PME selected four implementation partners to respond to their RFP. One implementation partner stood out, as they addressed some of the key assumptions PME included in the RFP:

- *PME included Marketing Cloud Account Engagement (Pardot) as an assumption for their marketing automation as it is mentioned to be ideal for B2B companies. However, one implementation partner mentioned in their response that – due to the fact that PME's roadmap includes an e-commerce store – PME should consider Marketing Cloud as it may be a better fit with Commerce Cloud.*

- *Middleware: PME has an existing middleware solution that connects their backend ERP system and current CRM system. Cary – PME's tech owner – assumed the current middleware was fit for modern enterprise architecture. An implementation partner pointed to several risks with using the legacy middleware – for one, it does not support* **event-driven architecture***.*

In addition to addressing the key assumptions, one implementation partner highlighted the need for a strategy for handling customer data. As Salesforce offers multiple clouds beyond the core (Sales Cloud and Service Cloud), multiple Salesforce clouds will hold information about customers. Managing the flow of customer data between systems will require effort to set up – either by leveraging provided connectors or by custom integration development. PME could also consider implementing a **Customer Data Platform (CDP)** *to act as a customer master data management solution. The implementation partner recommends that PME discusses this with its enterprise and data architects.*

Having received and evaluated the responses to your RFP, it's time to move on to the contracting phase.

Contracting with implementation partners

After you have put in the effort to follow the steps in the selection process, you want to make sure your project will not be hindered by any legal squabbles as your project progresses. With careful thought, you and your implementation partner can make the contract be a guiding star for your project rather than an administrative overhead.

Let's explore some of the documents you may come across in this contracting stage:

- **Statement of work (SoW):** This contract document describes the scope, pricing, invoice schedule, collaboration method, and how the delivered work is accepted in the project or engagement.

- **Purchase order (PO):** Following a co-signed SoW, a PO may be issued by your company and sent to your chosen implementation partner. It is used for accounting purposes as your implementation partner can state the PO number as a reference on the invoices they issue as they deliver work according to the contract.

- **Master services agreement (MSA):** Complements and supersedes other contract documents (SoWs and POs).

- **Terms and conditions (T&Cs):** Your company may have general T&Cs that are rarely negotiable on a per-engagement basis. Your legal counsel will know and guide you.

Important note

Connect with your procurement partner as well as your legal counsel to understand which structure your company uses. If your company does not have any in-house legal resources, it is highly recommended you engage with external counsel to assist you in drafting the contract for your Salesforce project and negotiating the terms.

Regardless of which legal document you end up working with, you *need to describe* – in addition to other legal and commercial terms – the elements in the proceeding table:

Typical contract clause descriptions per contract/engagement type:		
Contract/engagement type	**Fixed-price project**	**Time and material (T&M) engagement**
Delivery methodology	Pure waterfall.	Pure agile.
Scope (what the implementation partner is obligated to deliver)	Defined in the **Business Requirements Document (BRD)**.	High-level product vision. Ensuring suitably competent resources are assigned to the time agreed upon. This may be more or less detailed. It may be split by role (technical architect, developer, business analyst), by seniority/experience level, or by whether the resource is available onshore or offshore.
Price	Price per project milestone.	Rate card per resource role.

Invoice schedule	After project milestones are met.	On a monthly basis per reported billable hours.
Project management responsible	Supplier (implementation partner).	Customer.
Collaboration method and required client resources	Client-side resources availability at project milestones to assess, accept, or reject delivery.	Client-side resources available throughout the engagement to provide requirements and accept delivery of work.
Acceptance criteria of delivery of scope	In UAT and at go-live: The number of identified bugs at various criticality levels, and resolution time.	Work is continuously accepted in sprints/iterations.
Common appendices	BRD. SLA for bug fixing in UAT and hypercare.	Not applicable.

Table 4.2 – Typical contract element descriptions per contract type

As highlighted in *Table 4.2*, it is crucial to note how the elements' descriptions typically vary depending on your chosen delivery methodology.

> **Important note**
>
> *Table 4.2* describes the two pure delivery methodologies. In *Chapter 3, Determining How to Deliver Your Salesforce Project*, we deep-dived into these methodologies but also highlighted the common hybrid agile delivery methodology. You may have chosen a form of hybrid agile as the delivery methodology for your Salesforce project. If so, be sure to consider and adjust the contract elements accordingly – and be sure they don't conflict with each other.
>
> No matter whether the title of your contract or your project is agile, waterfall, or something else, be sure to align with your implementation partner and ask your lawyer to understand what each party's expectations are and *clearly describe* them for each element/term: fixed price or T&M, delivery, acceptance, timeline, scope, and so on.

Iterating in the pre-development phase

As mentioned previously, although the chapters of this book are laid out in a linear manner, you will iterate through the activities and decisions within them before you are ready to contract with Salesforce and your chosen implementation partner. Perhaps when presenting your Salesforce project's business case to senior leadership, you don't gain approval for the budget requested, or the business case does not live up to the criteria for your company's investment board. If this is the case, be sure to capture the reasons, so you know what to improve or modify. Then, go back to *Chapter 2, Defining the Nature of Your Salesforce Project*, to redefine the scope, considering dependencies between the capabilities.

Once you have gone through that iteration, continue through the cycle – the chapters in the pre-development phase – update your RAID log and business case, and present again.

Summary

In this chapter, you have summarized your Salesforce project in a business case containing a financial forecast based on the estimated revenue and cost impact of your Salesforce project. You have determined your partner model and target Salesforce CoE operating model and presented your business case to senior leadership to secure funding for your Salesforce project and have engaged with Salesforce to get a quote for your required licenses, and finally, you engaged with implementation partners, went through the process of selecting one, and contracted with your chosen implementation partner. Concurrently with selecting your implementation partner, you began establishing your Salesforce CoE by allocating resources – internal and external.

Up until now, you have created a vision for your Salesforce project – the *why* – as well as the nature of it – the *what*. You have determined how to deliver your Salesforce project – the *how* – and now you have forecasted *how much* your Salesforce project will cost and the financial impact it will have on your company by increasing revenue and reducing operational costs.

You have now completed all of the activities in the pre-development phase of your Salesforce implementation. Continue to the next chapter, which covers common issues to avoid in the pre-development phase and provides a checklist to evaluate the state of your Salesforce project.

Common Issues to Avoid in the Pre-Development Phase

The previous chapters covered a lot of activities in the pre-development phase: creating a vision for your Salesforce project, defining the business and technical scope, determining how to deliver it, building and presenting a business case to secure funding, and engaging with Salesforce and implementation partners.

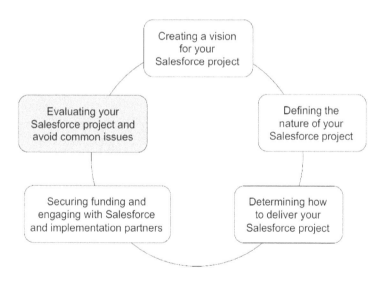

Figure 5.1 – Activities in the pre-development phase

In this chapter, we will discuss common issues often faced in the pre-development phase of a Salesforce project. We will also cover strategies to prevent or mitigate the issues from arising altogether.

You will be introduced to **root cause analysis** to dissect the common issues and be able to effectively solve them. The end of the chapter has a set of **checklists** to evaluate the state of your Salesforce project as you finish the pre-development phase.

This chapter will cover the following main topics:

- Introducing issues and root cause analysis
- Common issues in the pre-development phase
- Evaluating the state of your Salesforce project

Let's go!

Introducing issues and root cause analysis

Before jumping into what the common issues are in the pre-development phase of a Salesforce project, let's get some definitions straight.

In project management terms, **issues** are the manifestation of unaddressed risks. Moreover, unaddressed issues prevent a project from maximizing its potential for success.

Issues have underlying causes. **Root cause analysis** is a simple and effective method of stating the hypotheses for possible underlying causes of an issue. An issue may be the symptom of a larger issue that needs to be addressed, rather than simply alleviating the symptom.

In the next section, we will explore some of the most common issues that prevent a Salesforce project in the pre-development phase from moving forward. For each issue, we will cover possible root causes of the issue and provide strategies for mitigation and prevention.

Common issues in the pre-development phase and their root causes

Let's get started. With the definitions set straight, we'll now look at some of the most common issues faced in the pre-development phase, illustrated with a **fishbone diagram**:

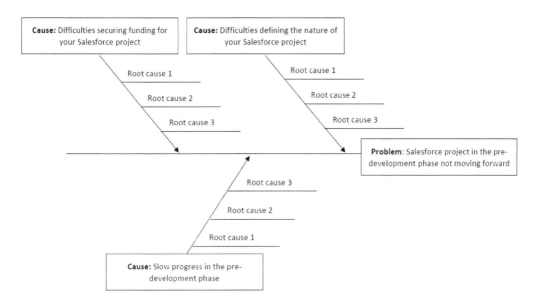

Figure 5.2 – Fishbone diagram of common issues in the pre-development phase

Next, we'll discuss the possible root causes and strategies for mitigation and prevention *for each* of the common issues in the fishbone diagram.

Difficulties in securing funding for your Salesforce project

In the following table, you'll find possible root causes for why you might be facing difficulties securing funding – along with strategies for mitigation if you are already facing that issue. If you have yet to try to secure funding for your Salesforce project, you can read along to know what to be mindful of – and then recap *Chapter 4, Securing Funding and Engaging with Salesforce and Implementation Partners*.

Possible root cause	Strategies for mitigation and prevention
The financial impact – ROI and payback – does not meet the criteria set by your company's investment board	Consider changing the phasing of your project. Create a scope-geography roll-out matrix to visualize how Salesforce will cover certain geographies and scopes. Update your business case accordingly – a smaller and faster initial implementation and roll-out investment will accelerate time to value.
Your business case is incoherent to the stakeholders	Work with your finance business partner to understand the required format.

The business capabilities in scope are not tied to strategic business priorities	You need to work with your executive sponsor to understand the overall strategy of your organization. Revisit *Chapter 1, Creating a Vision for Your Salesforce Project*, and progress through the pre-development phase activities.
Lack of involvement from your executive sponsor	Realize that having an engaged executive sponsor is a *prerequisite to the success of your Salesforce project*. Set up a meeting with your executive sponsor and re-align on the following: • Alignment on the vision for the Salesforce project • The expected involvement of your executive sponsor in the project • Discuss the challenges you are facing to secure funding for the Salesforce project – ask for guidance on how to proceed • Execute the provided guidance
Unsure what targets to set for your KPIs	A general rule of thumb is if you are a medium or large company and your company is currently relying on spreadsheets and email to manage your sales and customer service functions, then investing in a CRM – with adequate considerations and change management – will be a fruitful investment. If your project's scope is an extension of your current org – such as Field Service, Commerce, Marketing Cloud, or CPQ (Configure, Price, Quote)– you should engage a domain expert or architect to understand your industry benchmark improvement when implementing such solutions.

Table 5.1 – Possible root causes for difficulties in securing funding for your Salesforce project

Let's jump to the next common issue in the pre-development phase.

Difficulties in defining the nature of your Salesforce project

In *Table 5.2*, you'll find possible root causes for why you might struggle to define the nature of your Salesforce project – along with strategies for mitigation if you are already facing that issue. If you are yet to define the nature of your Salesforce project, you can review the table to know what to be mindful of – and then recap *Chapter 1, Creating a Vision for Your Salesforce Project*, and *Chapter 2, Defining the Nature of Your Salesforce Project*.

Possible root cause	Strategies for mitigation and prevention
Unclear project vision and internal alignment	Revisit *Chapter 1, Creating a Vision for Your Salesforce Project*, and engage with your executive sponsor and wider business stakeholders to create an aligned project vision.
Not having involved end users	Restart and reset. Not involving end users in the pre-development phase of your Salesforce project is a recipe for disaster. Be sure to at least hear their concerns and pain points with their current processes. Incorporate these into the scope of your project and consider any changes to the technical scope and the overall business case.
Insufficient technical resources in your Salesforce taskforce to determine the hypothetical technical scope	If your company doesn't have in-house technical resources, engage an external Salesforce architect to draft the hypothetical technical scope based on the business capabilities to include in the scope. Then update your business case accordingly.

Table 5.2 – Possible root causes for difficulties in determining the nature of your Salesforce project

> **Tip!**
>
> Consult the executive sponsor of your Salesforce project to understand which capabilities are considered strategic within your company – providing your own thoughts based on the findings from your workshops with end users.

Let's jump to the next common issue in the pre-development phase.

Slow progress in the pre-development phase

In *Table 5.3*, you'll find possible root causes for why you may experience slow progress for your Salesforce project in the pre-development phase – along with strategies for mitigation if you are already facing that issue:

Possible root cause	Strategies for mitigation and prevention
You got stuck analyzing *Chapter 1*, *Creating a Vision for Your Salesforce Project*, and *Chapter 2*, *Defining the Nature of Your Salesforce Project*	Realize that the analysis and decisions made until this point – while important – are not final. Keep an open mind and update your assumptions as you progress.
You are challenged by having to iterate multiple times based on new information	While you should iterate and update your business case based on new findings and feedback, don't let perfect get in the way of good and done.
Your project is not a greenfield implementation and you are merely extending with another cloud or feature	In this case, you will iterate through the steps in the pre-development phase much more quickly. Consider engaging a implementation partner to do the "discovery" with you to shape the project vision, goals, scope, and business case. While this is certainly an option, this approach works best when internal organizational alignment and a defined project portfolio management framework are set up. We'll discuss program management and continuous improvement in *Chapter 12*, *Evolving Your Salesforce Org and DevOps Capabilities*.

Table 5.3 – Possible root causes for slow progress in the pre-development phase

Next, let's evaluate your Salesforce project.

Evaluating the state of your Salesforce project

We have covered many topics, domains, and activities in this pre-development phase of a Salesforce project. The purpose is to enable you to deliver successful Salesforce projects.

If your project seems to be stuck, or you are not pleased with the progress, the following section will help evaluate the state of your Salesforce project. For each item you cannot check, go back to the corresponding chapters and complete the activities to proceed with your Salesforce project.

Checklist for your Salesforce project in the pre-development phase

Embarking on the pre-development phase, you have made the following foundational steps:

- Enlisted an executive sponsor for your Salesforce project
- Created a Salesforce taskforce to drive the activities in the pre-development phase

Next, you have completed these activities to understand your company's strategic situation:

- Understood how your company is organized
- Understood who your customers are
- Understood the products and services your company sells
- Understood your company's competitive situation
- Understood *why now* is the right time for your Salesforce project
- Created a *vision* for your Salesforce project – aligned with your executive sponsor and key stakeholders

Next, you have completed these activities to define the nature of your Salesforce project:

- Understood the business capabilities in the scope
  - People, processes, systems, and data
- Involved end users
- Envisioned future capabilities
- Determined the high-level technical scope
  - Have included technical enablers and non-functional requirements
- Started building your **risks, assumptions, issues, and decisions (RAID)** log

Next, you have completed the following activities to decide how to deliver your Salesforce project:

- Understood whether you are preparing a Salesforce project or program
- Chosen the delivery methodology for your Salesforce project
  - If hybrid, you have articulated and described how you define your methodology
- Considered your project phasing approach (geography and scope)
- Determined your change management and communication strategy

- Considered the degrees of change impact
- Considered the timing of your change management activities

- Envisioned your Salesforce **Center of Excellence (CoE)**

  - Created a charter
  - Determined the structure and responsibilities of the CoE
  - Created a roadmap for your Salesforce program

Next, you have completed these steps to secure funding for your Salesforce project:

- Defined KPIs and targets for them
- Created a financial forecast

  - Estimated the revenue impact
  - Estimated the cost impact

    - Operational costs
    - License costs
    - Implementation and roll-out
    - Determined your Salesforce CoE operating model

  - Calculated the **profit and loss (P&L)** forecast, ROI, and payback of your Salesforce project

- Presented your Salesforce project business case to senior management

After securing funding for your project, you have done the following:

- Requested a quote from Salesforce
- Defined your process for selecting an implementation partner
- Evaluated and selected an implementation partner
- Contracted with an implementation partner

  - Been mindful in describing the responsibilities of each party in the contract

- Established your Salesforce CoE

Throughout each chapter in your pre-development phase journey, you may have found yourself having to iterate. Take a step back, change some elements, and then keep going.

Summary

In this chapter, you have assessed the state of your Salesforce project in the pre-development phase by going through checklists. You have also become familiar with the common issues often faced in the pre-development phase and learned how to mitigate or prevent them from arising.

With your Salesforce project in great shape, you are ready to move on to the next step in your journey, the development phase, starting with *Chapter 6, Detailing the Scope and Design of Your Initial Release*, where we will co-create your business processes, write user stories, and create the solution design for your initial Salesforce release.

Part 2: The Development Phase

This part will teach you how to further define the scope of your Salesforce project and to understand the solution design for it. We'll also take you through the development process and testing activities, and discuss common issues so you can avoid or mitigate them.

This part has the following chapters:

6

Detailing the Scope and Design of Your Initial Release

Following the hard work, you did in *Part 1*, *The Pre-Development Phase*, this first chapter in the **development phase** of your Salesforce project will guide you through key activities and milestones. Next, we cover how you break down the high-level scope you defined in the pre-development phase into **user stories**. Then, we will cover how to determine an appropriate level of upfront architecture, and what **Salesforce architecture artifacts** you should create. Finally, we'll discuss how to understand user story solution design and illustrate your **target solution architecture**.

The chapter provides best practices for creating user stories to maximize the chance of developing the *right* solution for your business. We describe which solution design artifacts you should create and maintain throughout the development phase of your Salesforce project, and beyond!

In this chapter, we'll cover the following main topics:

- Overview of your Salesforce project in the development phase
- Detailing the scope of your initial release
- Creating the solution design for your initial release

Let's go!

Overview of the development phase of your Salesforce project

This phase of your Salesforce project is the one architects, business analysts, project managers, scrum masters, and developers are most familiar with. The focus is on detailing requirements, designing solutions, and building and testing them.

Let's take a look at the milestones of the phase.

Key activities and milestones of the development phase

The phase starts at **project kickoff** where the project manager, business leaders, architects, and project sponsor introduce the project vision and nature to the project team members onboarded at this point.

After the project kickoff, the next key milestones in the development phase are as follows:

- Detailing the scope of your release
- Completing the solution design of your release

> **Note**
> The activities to reach the first two milestones are sometimes referred to as the **prepare phase**, as the project is preparing to start building the solution.

- Establishing your team's development process
- Completing the build and testing of your release
- Creating your data migration plan
- Detailing your initial roll-out, training, and change management plan

> **Tip**
> If you have chosen a pure agile delivery methodology for your project, you will find inspiration for your setup in *Chapter 12, Evolving Your Salesforce Org and DevOps Capabilities.*

Having established *what* the overall milestones and activities consist of in the development phase, let's look at *how long* your development phase may take.

Duration of the development phase

Depending on the following main factors, the development phase of your Salesforce project may take anywhere between *a few weeks to many months* to complete:

- The scope of your Salesforce project
- Your chosen delivery methodology:
  - **Pure agile**: You'll be releasing working software to production on demand when features are developed
  - **Pure waterfall**: You deploy your release at the end of the project when the release has passed user acceptance testing
  - **Hybrid**: You may either be releasing as per the pure agile or pure waterfall delivery methodology

- Allocated and available resources:

  - **Business resources**: This includes a dedicated **product owner** (**PO**) to own and gather the requirements, and validate the delivery of functionality. The availability, dedication, experience, and competence of the PO all have a major impact on the project.

  - **Technical resources**: These are architects, functional Salesforce consultants, Salesforce developers, and QA specialists (testers), as well as other technical resources that may need to make changes to the systems to be integrated with Salesforce. The coordination and timing of the availability of the required resources is a key task to optimize the project delivery.

Tip

Unless your project is a greenfield implementation – meaning it's the first time your company will use Salesforce – you should *limit the prepare phase* to a maximum of a few months for larger projects (for example, new clouds such as Field Service or major products such as CPQ) to ensure requirements input and target processes are still relevant.

For the sake of simplicity, this section of the book will focus on a *hybrid agile delivery methodology* in which some activities are done to detail the scope of the Salesforce project and create upfront architecture before starting the build phase.

Important note

The milestones and activities after project kickoff will either happen in sprints (pure agile), stages (waterfall), or a mix (hybrid agile).

To refresh the differences, please see *Figure 3.2* in *Chapter 3, Determining How to Deliver Your Salesforce Project*.

Let's see who will be overseeing your Salesforce project.

The role of your Salesforce CoE in the development phase

Recall the *Salesforce taskforce* that you were encouraged to establish to drive the activities in the pre-development phase. It included resources from three key areas of your company: business, technical, and project management office. *You* likely reside in one of these functions.

Your Salesforce taskforce has established the capabilities and Salesforce clouds in the scope of your project – and secured a budget for at least the initial release. Your Salesforce taskforce has also contracted with an implementation partner to support delivering the initial release.

Your Salesforce taskforce shouldn't be shut down after the pre-development phase, rather it should evolve into your Salesforce **Center of Excellence** (**CoE**). The members of your Salesforce CoE are well suited for *governing the delivery of your initial release*.

In *Chapter 3, Determining How to Deliver Your Salesforce Project*, you envisioned the structure and responsibilities of your Salesforce CoE. Now it is time to execute and establish it.

The next activity – in the hybrid delivery methodology – is to break down the business capabilities into smaller backlog items that your implementation partner's developers and functional consultants can consume and deliver working solutions for.

Detailing the scope of your initial release

Your project's next endeavor consists of translating the vision and capabilities – business processes – into product backlog items. Breaking down the business capabilities into smaller backlog items requires you to involve the end users of your Salesforce solution through a series of workshops.

In *Chapter 3, Determining How to Deliver Your Salesforce Project*, we discussed the three overall delivery methodologies to choose from – *pure waterfall*, *hybrid agile*, and *pure agile*.

As we are covering a hybrid agile delivery methodology in this section, let's consider the activities to prepare for development:

- An effort is made to establish foundational architecture including – but not limited to – determining a Salesforce org strategy, integration, data migration, data backup and archival, and reporting strategy, as well as a development and environment strategy

- Detail the business processes (the customer and user journeys), in the scope of your project, and create a solution design for it

> Tip
>
> In hybrid agile – after establishing the foundational architecture – you can opt to detail all processes in the scope of your project – along with solutions and estimates. Alternatively, you can decide to detail a sufficient number of backlog items to fill the first couple of sprints, and, concurrently with the sprint development team, dedicate resources to continuous backlog refinement in a parallel design stream.

Let's get started detailing your company's future business processes.

Co-creating your future business processes

You may recall the activities in *Chapter 2, Defining the Nature of Your Salesforce Project*, where you facilitated workshops to determine current capabilities, their processes, and the current pain points and suggestions from the stakeholders.

Whereas the purpose of those workshops was to determine and understand current processes, it is the purpose of the workshops in the development phase to create future processes supported by Salesforce.

Plan and facilitate workshops for each of the capabilities – processes – in the scope of your initial release. Review the workshop structure proposed in *Chapter 2, Defining the Nature of Your Salesforce Project*, but follow this proposed agenda:

- **Welcome and introduction**: The facilitator welcomes the participants and introduces the agenda and goal, and the participants introduce themselves:

  - **Workshop goal**: Share the vision statement you crafted in *Chapter 1, Creating a Vision for Your Salesforce Project*

    - You are seeking your participant's valuable input to shape your company's future CRM system

- **Exercises**: Divide the overall workshop goal into smaller exercises divided by adequate breaks:

  - **First exercise**: Review the current process as you defined in *Chapter 2, Defining the Nature of Your Salesforce Project*, and review the pain points and suggestions for improvement:

    - This approach – while valuable in itself – has the added benefit of creating a sense of urgency and sparking the desire for change

  - **Second exercise**: It's time to start with a clean slate and begin co-creating the to-be processes. This can be daunting – staring at the blank canvas as a group. Here are a few methods to help you get started:

    - Determine the start and finish of the overall business process by briefly discussing what has happened before the starting point of the process, and what will happen after the finishing point.

    - State the over arching phases in the process: for example, creation, assignment, execution, and post-execution.

    - Create **swim lanes** by determining what **personas** are part of the business process – you may reuse the personas you determined for your current process.

Tip

Keep the goal and focus of the workshop in mind, and guide your participants so they don't get stuck overanalyzing minute details of each process step.

    - As mentioned earlier, pay *extreme attention to variations*. If your solution is to be used by multiple regions and/or lines of business, be sure to include stakeholders from each – where practically possible – if it is the ambition of the Salesforce project to harmonize business processes.

> **Tip**
> Ensure you align with your implementation partner before designing the workshops to determine each party's roles and responsibilities to maximize the output of the workshops. If you are working with an implementation partner, they should be able to provide experienced business analysts and/or architects to facilitate and drive the workshops.

Depending on the scope and nature of the business capabilities in the scope of your Salesforce project, an experienced **Salesforce business analyst** may suffice to support your business stakeholders in co-creating your future business processes supported by Salesforce. For greenfield implementations, and for more complex implementations, be sure to include a **Salesforce solution architect** in the workshops. Their input and experience will make sure the proposed processes make sense from a Salesforce perspective.

Let's look at the next steps after each workshop.

Crafting user stories

Following each workshop, your PO or your implementation partner will use the co-created future business processes to create backlog items, also known as **user stories** for each epic or business process.

A user story should be an easy-to-understand sentence describing – from a *user*'s perspective – *what* they want to be able to do and *why*. Let's break it down:

- **Persona**: They may be internal or external persons who are part of the business process:

  - Internal persona examples: An employee in a specific function

  - External persona examples: A customer, partner, or supplier

- **Need**: What does the persona want to be able to do in the process step?

- **Purpose**: What business outcome does this achieve?

  - Be mindful that the purpose should tie back to the overall business objectives of the project. The purpose of a user story *is not* merely to be able to proceed to the next step in the business process.

A user story should be written for each step in your business process.

Some considerations to keep in mind when writing user stories are as follows:

- They should not be written in Salesforce-specific language, rather it should be technology and system agnostic

- They should focus on the business need, problem, and desired business outcomes, and not describe a desired solution

Sticking to the preceding advice will ensure the expert consultants at your implementation partner seek to find the best solution for your business.

> **Important note**
>
> Ensure that a PO is allocated to own each domain of your Salesforce solution. If a single business stakeholder does not represent all business domains in the scope of your solution, you can consider either promoting one to own all domains or allocating multiple POs responsible for different domains. Be vigilant about this decision, as you will need to have an assigned **chief PO** to make final decisions in case requirements from two POs are contradictive or overlapping.

Reviewing and detailing user stories

Your PO *reviews all user stories* for the business processes they own.

While the goal of the workshop was to ensure that the end-to-end business process for the capability is sound, lean, and logical, and one which will drive great user and customer experience, the goal at this stage is to ensure that the details and context of each user story are aligned and documented. As such, the PO should do the following:

- **Review** the proposed user stories' *user*, *need*, and *purpose*, and make required modifications.

- Add **further context** to help other business stakeholders understand the situation from the persona's perspective in the process step. This may include additional descriptions, documents, screenshots, lists, or other media.

- While it may not be strictly required in pure agile, adding *short and simple* **acceptance criteria** statements to user stories is strongly recommended for three key reasons:

  - It enables each user story to be tied to the desired business outcomes of the epics included in your Salesforce project.

  - It provides clarity for other business stakeholders as to what the scope is for the user stories – and what is not. If it's not a part of the acceptance criteria, it's not in the scope.

  - It guides the implementation partner's solution architect and developers to understand the expected outcome of the user story.

The value in having your PO – and possibly a wider business stakeholder group – review and detail user stories are plentiful:

- **Consolidates process ownership**: Having a dedicated PO who owns a business process is essential. It is a prerequisite, especially if variations exist in your current business processes and your future processes need to be harmonized amongst geographies and lines of business as part of your Salesforce project.

- **Lowers cost by "shifting left"**: Ensuring business alignment and clarity early will save you money – conversely, the further down the development pipeline you fix an issue, the greater the cost:

  - If a piece of functionality is rejected later in UAT or, worse yet, not adopted by end users in production, it will carry a huge cost to rectify due to all the steps and people involved in getting a piece of functionality through the pipeline (specification, solutioning, testing, and deploying).

- **Embeds change management** in the development process:

  - Recruiting and involving change agents in process co-creation workshops as well as validating acceptance criteria is an excellent strategy to continuously lay the groundwork for change management

  - Including end users early and often reduces the change management effort and pressure in later stages, such as UAT and training

  - Tightly related to the previous point, acceptance criteria can – when aggregated – form the basis for user acceptance testing

> Tip
>
> Capture user stories in fit-for-purpose tools. **Atlassian JIRA** and **Microsoft Azure DevOps** are the most commonly used tools for Salesforce projects. Simpler alternatives exist, including **Trello**, although with less functionality and suitability for enterprise Salesforce projects.

Organizations with less experience with agile or hybrid agile methodologies often struggle with the mindset change from traditional waterfall projects. Being used to creating a giant **business requirements document** is fundamentally different from creating user stories with acceptance criteria.

While most project team members get the hang of it after some time and practice – and guidance from implementation partners – investing in training courses or self-directed learning accelerates the process and delivers higher-quality user stories with better solutions and adoption rates.

> **Tip**
>
> For more information, you can refer to the *Salesforce Business Analyst Handbook* (`https://www.amazon.com/Salesforce-Business-Analyst-Handbook-techniques/dp/1801813426`) by Srini Munagavalasa.

As user stories are reviewed and detailed, let's begin solutioning.

Creating the solution design for your initial release

The user stories created for the business processes collectively define the scope of your Salesforce project. Although the solutions for each user story collectively make up your technical solution, you will benefit from *considering your overall and end-to-end architecture* to some degree before starting development.

Let's start by looking at the architecture artifacts and components you should have create and maintain as part of your project – and beyond!

Understanding Salesforce architecture artifacts

Governing the documentation for your solution is a key responsibility of the Salesforce CoE.

While the source of the documentation components may be your implementation partner, it is your Salesforce CoE's responsibility to ensure that the solution documentation is available, up to date, and understandable for the intended audiences – both business and technical stakeholders.

Let's take a look at the key artifacts for your Salesforce solution:

- **Business capability map**: You created this in *Chapter 2, Defining the Nature of Your Salesforce Project*, but circumstances may have changed since then, so go ahead and update your business capability map to reflect the scope of your Salesforce project. This diagram will introduce the audience to the overall scope of your solution.

- **Business processes diagrams**: You made these earlier in this chapter in the *Co-creating your future business processes* section.

- **Actor and licenses diagram**: Show which actors (persona) will use or interact with your Salesforce solution. Highlight which actors are internal and which are external, as well as which Salesforce user license their users will have.

- **Target system landscape**: This diagram illustrates the Salesforce clouds and AppExchange apps used within Salesforce, and critically, how Salesforce is connected to other systems in your enterprise system landscape. Each interface in the scope of your solution should be labeled, as well as the direction.

- **Integration**: To supplement the system landscape diagram, you should have a table describing the integrations in the scope of your solutions, including the following:

- **Direction**: What system is the source and target for each interface, or whether it is a bi-directional interface.

- **Layer**: Whether the integration is a data, process, or virtual integration.

- **Timing**: Whether the integration is synchronous (blocking) or asynchronous (non-blocking or queued).

- **Pattern**: The pattern for the integration. **Remote-process invocation fire-and-forget (RPI-FF)**, **batch data synchronization (BDS)**, and **data virtualization (DV)** are examples of standard integration patterns on the Salesforce platform.

- **Authentication**: How each interface will gain access to the target system.

- **Data model**: You critically need to create and maintain the data model for your Salesforce solution. The data model is a graphical representation of the objects and fields used in your Salesforce solution. Many objects and fields come **out of the box (OOTB)** with your Salesforce licenses, but you will extend that data model with your own custom fields and likely some custom objects as well:

  - **Object list**: To complement your data model, it is good practice to maintain an object list detailing the use of each object. Include the following attributes for each object: object name, standard/custom, what entities the object will hold, record types, record owner, org-wide default (internal and external), and volume estimate.

  - **Data security matrix**: State the object permissions (read, create, edit, and delete) for each object, profile, and persona.

  - **Data dictionary**: It is good practice to create and maintain a dictionary for all the fields used in your solution. Your data dictionary is valuable for integration architects and developers, as well as business users, who will find it useful when creating reports.

- **Sharing model**: You should describe what mechanisms will be used to grant the right users the right access to the right records. The **role hierarchy** is an integral part of your sharing model and is complemented by a vast number of options to share data.

- **Data backup and archival**: You will benefit later by already considering and deciding how to back up your data in Salesforce in case you need to restore it – and how (which tool) you plan to use for it. Determining your data archival policy and method by object should also be done at this stage to prevent you from hitting storage limits unexpectedly.

- **Environment and development strategy**: The environment overview diagram should describe which Salesforce development environments (sandboxes) will be used in the development and deployment of your solution, as well as the tools used to migrate metadata and data between environments.

While the input for creating some of the architecture artifacts is the output of the workshops where you co-created future processes, determining other aspects of your architecture requires separate workshops with a mix of technical and business stakeholders. These aspects include system landscape, integration, sharing model, environment, and development strategy.

> **Tip!**
>
> To understand in depth the technical architecture artifacts briefly described, you should read the *Becoming a Salesforce Certified Technical Architect* book, written by a Salesforce **Certified Technical Architect (CTA)**, Tameem Bahri.

If your company has a diagramming framework, use that. If not, you can either use Salesforce's diagramming framework (`https://architect.salesforce.com/diagrams#framework`) yourself or require your implementation partner to adopt it.

> **Tip**
>
> Salesforce provides templates for many of the Salesforce cloud data models and **industry reference architecture templates**, which are freely available to use in **LucidChart** or MS PowerPoint format at `https://architect.salesforce.com/diagrams#design-patterns`.

Let's move on to discuss what level of *initial* architecture makes sense to establish for your Salesforce project.

Determining the appropriate level of upfront architecture

If you have chosen hybrid agile as the delivery methodology for your Salesforce project, you have a key decision to make:

"How much effort should my project allocate to architect my Salesforce solution before starting development?"

You need to consider *two levels* of solution design: overall technical architecture, which we described in the previous sections, and the user story level. You should expect your implementation partner to provide solution design at both levels.

Let's jump right in and look at some of the benefits of creating architecture upfront:

- Ensures the end-to-end architecture is cohesive and comprehensive and establishes the *technical foundation* to ensure your Salesforce solution can be secure, performant, and scalable

- Identifies technical issues and gaps that can be handled and mitigated proactively

- Guides the development team and provides guard rails

- Increases **development velocity** through higher **sprint efficiency** because your team members don't have to spend time creating solution designs in a sprint

If you don't establish any architecture before starting development, you simply won't have those benefits – and critically, you won't know the gaps and risks, and as a result, you can expect to hit many obstacles when building your solution.

The goal of your Salesforce project's prepare phase is to establish stakeholder alignment and to create just enough **architectural runway** for the development team to be able to start sprinting.

> **Important note**
>
> You should expect your implementation partner to make a recommendation for what level of architecture to determine upfront for your specific project. The contract with your implementation partner may describe how the project should be delivered, so be sure both parties are aligned on what methodology to follow and what it means.

Let's move on to the next level of solution design.

Creating user story solution designs

After you have worked with your implementation partner to write user stories and made sure the acceptance criteria make it possible to propose solutions for them, your implementation partner can begin to provide solution designs for the user stories.

As we mentioned in the *Detailing the scope of your initial release* section, you can opt to have all business processes broken into user stories upfront or have a design stream that continuously prepares user stories for the development team.

The optimal approach depends on a number of factors, including the breadth of your Salesforce project's scope – the number of different business capabilities, your organization's familiarity with Salesforce, and hence its ability to articulate and detail user stories and comprehend the proposed Salesforce solutions.

As a *rough* rule of thumb, if your solution was – in the pre-development phase – initially estimated to require less than 5 sprints (10 weeks) for a small development team to build, you may consider breaking down all the business processes to user stories, and have solutions proposed and estimated prior to beginning development. If you choose to do so, you should still be mindful that some things may change as you start building your solution – your PO may be inspired or receive new input from end users.

Evaluating user story solution designs

Your implementation partner should propose solution designs for your user stories.

This section will provide you with methods to understand, reflect on, and challenge the proposed solution designs without requiring any technical competencies on your part.

For any user story – and overall business process – the following considerations should be made when choosing a solution:

1. Is the functionality available OOTB with your current licenses?

2. Investigate whether a free solution exists on the **AppExchange**.

3. Buy. Here, you may have two options:

    A. Check whether a solution is available with a Salesforce add-on product:

        • This way, you will benefit from further enhancement and be able to grow with the platform rather than having to continuously invest to ensure alignment (and regression test)

    B. Investigate available paid AppExchange products.

4. Build your own solution:

    A. First, investigate to what extent **configuration solutions** (no code) can meet the requirements. Salesforce prides itself on its extensive no-code capabilities, and you should expect the vast majority of proposed solutions to be configuration:

        • If a user story can only partially be solved with configuration solutions, consider – together with the PO – whether the user story and its acceptance criteria can be tweaked to allow for configuration solution

    B. Next, consider a **custom solution** (code):

        • While not always true, more often than not, custom solutions require more effort, cost more, and require more maintenance. Hence, you should only consider custom solutions as a last resort.

Another key aspect of evaluating solution designs is their collective level of required **maintenance**. If you have chosen to detail all user stories and have solution designs proposed for them, you can – *and should* – already now evaluate the required maintenance of the proposed solutions. If you plan to continuously refine your product backlog for development, the level of required maintenance is a factor you can – to some degree – influence by having the right governance mechanisms in place for your Salesforce project. We'll cover this in more detail in *Chapter 7, Building and Testing Your Initial Release*.

Tip

Ask your implementation partner to specify the maintenance required for each user story's proposed solution. Summarize this and compare it to your initial hypothesis and target Salesforce CoE operating model.

In addition to this, you can calculate a tangible metric to understand how custom the proposed overall solution is. Summarize by a count of user stories the percentage for which a customized solution (code) is proposed – knowing whether your implementation partner is planning to use programmatic development for 5%, 10%, or 20%+ of user stories makes an enormous difference in terms of development effort and required maintenance. Track this metric as the solution design process continues.

Let's see the split of proposed user story solution types for the Salesforce solution of our scenario company, **Packt Manufacturing Equipment (PME)**:

Solution type	Number of user stories	Share of total user stories
Configuration	109	84%
Data	7	5%
Customization	13	10%
Total	129	100%

Table 6.1 – PME's split of proposed user story solution types

In PME's Salesforce project's prepare phase, project members from both PME and its implementation partner have decided to break down all user stories upfront before initiating the development phase. PME's leadership wants to ensure a clear target solution architecture is established and evaluated. They also believe - based on input from their implementation partner - the project scope can be delivered within four development sprints.

The project team has been taking note of the types of solutions proposed for each user story and created the summary presented in *Table 6.1*.

Analyzing the drivers for the split, they found the following insights:

- **Configuration**: This includes both solutions leveraging OOTB features and solutions where free AppExchange solutions are proposed, and solutions where declarative tools are used to build functionality, including **Salesforce Flows**.

- **Data**: Some parts of the Salesforce platform features rely on data, as opposed to **metadata**, to create solutions. For PME, these include **Products**, **Price Books**, and **Price Book Entries**.

- **Customization**: Although PME intends to leverage declarative solutions for its integration requirements (**Enhanced External Services** and **Platform Events**), some customization solutions are necessary, including the **email-to-case handler** (**Apex class**).

If you find a great percentage of the proposed solutions for your Salesforce project are programmatic solutions, you should challenge both your implementation partner and your Salesforce PO:

Could the customization user stories be modified slightly to allow for more configuration solutions?

Illustrating your consolidated solution architecture

To communicate your future Salesforce solution at an abstraction level where project stakeholders can understand it, you should create a **target solution architecture diagram** (and maintain) for your Salesforce project. The diagram is sometimes referred to as a **functional architecture diagram**, but it is essentially the same thing. While your system landscape diagram looks similar, the solution architecture diagram adds more context, including key objects, and Salesforce key features and products.

Let's have a look at PME's target solution architecture in the following figure:

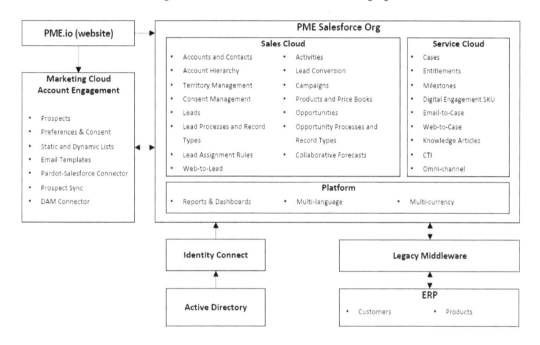

Figure 6.1 – PME's target solution architecture

The PME target solution focuses on **Salesforce Sales Cloud**, **Service Cloud**, and **Marketing Cloud Account Engagement** (formerly marketed as **Pardot**). The solution leverages OOTB functionality with limited key features being customized. The solution integrates with PME's website via **Web-to-Case** to allow customers to create cases by filling out a form. In addition to Web-to-Case, PME will utilize Pardot forms to collect leads from their website.

Although PME's implementation partner proposed to use a new modern middleware, such as **Mulesoft**, PME's leadership said it would not do so as part of the initial Salesforce implementation. They realize it needs to be done at some point and decided to put it on the roadmap.

To make user management efficient, PME will configure and integrate with Salesforce **Identity Connect** to provision and de-provision users and manage their access rights centrally.

Summary

You have made great progress throughout this chapter. You were introduced to the key milestones and activities in the development phase of your Salesforce project. Next, you learned how to break down your business processes into user stories with acceptance criteria. Then, you were introduced to the Salesforce architecture artifacts that are key for your Salesforce project. Finally, you learned how to evaluate and interpret the proposed solutions for your user stories and illustrate your target solution architecture.

Now that you Having established the foundational architectural runway and solution design for your Salesforce project, you are ready to move on to the next chapter of your Salesforce implementation, *Chapter 7, Building and Testing Your Initial Release.*

7

Building and Testing Your Initial Release

In this chapter, we will cover how to build and test your Salesforce solution for initial release. We will go through what your solution consists of under the hood, the available Salesforce development models, and sandboxes. Then we'll describe the agile team roles, the value of agile ceremonies, and how to effectively manage your development process. We'll cover the governance mechanisms you can, and should, consider putting in place to steer your Salesforce project and the changes that may occur along the way. We'll describe the various types of development phase testing you should be familiar with. Finally, we'll go through the steps to create your data migration plan.

You will learn about the Salesforce development life cycle and how to organize an agile team consisting of both internal and external members, how to facilitate agile ceremonies, and how to govern the delivery of your Salesforce project. You will also learn how to distinguish between different types of testing for your Salesforce solution. Then, we'll discuss ways of measuring progress. Finally, you will learn how to execute due diligence for data migration planning.

We will follow the Salesforce CoE of **Packt Manufacturing Equipment** (**PME**) as it oversees and collaborates with the company's implementation partner.

In this chapter, we'll cover the following main topics:

- Understanding the roles and responsibilities of your team members
- Planning your build phase
- Setting up your development model and environments
- Establishing your development process
- Governing your Salesforce project in the development phase
- Preparing your data migration plan

Let's go!

Understanding the roles and responsibilities of your team members

It is the people in your team that will make or break your Salesforce project. Ensure you invest in making sure roles and responsibilities are clear, processes are optimized to deliver value as effectively as possible, and that you have a learning and growth mindset to continuously improve your setup.

In the previous chapter, business analysts and architects interacted with your business and technical stakeholders to break down the overall scope of your Salesforce project into user stories and create the solution design artifacts.

In this chapter, we'll cover a hybrid agile delivery methodology, and as we are in the development phase, we'll go deeper into its agile aspects.

You will begin configuring and developing your Salesforce solution. Depending on your chosen delivery methodology, some project members may not yet have been onboarded – but they are about to be.

Let's start by looking at the *common* Salesforce project and agile team member *roles* and their *main responsibilities*:

- **The product owner (PO):**

  - Owns and manages the backlog of their agile team and prioritizes the backlog ahead of sprint planning. The PO is also responsible for stakeholder management and release management from a business perspective.

  - Is responsible for answering questions from functional consultants and developers to provide clarification and further details on user stories.

  - Is responsible for validating developed user stories against acceptance criteria.

  - Is responsible for consolidating feedback from stakeholders (together with the business analyst) at product demos, creating user stories, and communicating decisions back to stakeholders.

> Tip
>
> If you plan to develop in sprints – which is the underlying assumption for this chapter – make sure you allocate a *competent and empowered PO* to work in the agile development team continuously throughout your project. Your PO is instrumental in backlog refinement and in validating the user stories developed by the development team.

- **The scrum master:**

  - Facilitates agile ceremonies and coaches team members to understand their roles and responsibilities, processes, and agile ceremonies and values.

- Removes obstacles and blockers or escalates issues if they are unresolvable within the team.

- Should aim to shield the development team from meetings and administrative work, so they can focus on building solutions.

- Is a role typically fulfilled by your implementation partner. Many implementation partners have a development methodology they follow – as such, it makes sense for them to staff the scrum master role. If your contract with the implementation partner has any aspect of fixed price and timely delivery, this role must be external to ensure accountability.

- **The development team**: This consists of the people who perform the important low-level solution design, development, testing, and deployment of user stories. Each member of the development team can have one or more responsibilities, depending on their skill set and the nature and scope of your Salesforce project. The following are some of the common roles and responsibilities you will find in your Salesforce project's development team:

  - **The functional consultant**:

    - Is responsible for developing declarative solutions (point-and-click) on the Salesforce platform.

    - Can have expertise in one or more Salesforce products or subdomains of CRM. No functional consultant masters all clouds, so be sure that your team has the relevant knowledge and experience to match the scope of your Salesforce project.

    - Is a role typically fulfilled by your implementation partner, but if you have internal resources with Salesforce declarative development skills, you can opt to include them in the agile team. Again, be mindful of what the contract with your implementation partners states.

  - **The developer**:

    - Is responsible for developing programmatic (code) solutions on the Salesforce platform.

    - Can have expertise in one or more Salesforce products and/or coding languages. It is not common for a developer to master all products and technologies. For example, being a Marketing Cloud developer, coding AMPscript is vastly different from coding in Apex and building **Lightning web components** (**LWC**) on the core platform. Make sure your developers have the relevant knowledge and experience to match the scope of your Salesforce project.

    - Is a role typically fulfilled by your implementation partner, but if you have internal resources with suitable development skills, you can opt to include them in the agile team while being mindful of your contract with the implementation partner.

In a *Scrum* setup, the PO, scrum master, and development team are *collectively responsible* for ensuring the architecture is sound and the solution is of high quality and bug-free and for releasing, monitoring, and supporting change and adoption.

However, for enterprise projects – including Salesforce projects – it is crucial to have the following specialized roles as part of your project throughout the development and roll-out phases of your Salesforce project:

- **The Salesforce architect**:

  - In the hybrid agile delivery methodology, you allocate time and effort to create some degree of upfront solution architecture. What commonly happens is that circumstances and requirements change or assumptions turn out to not be right. In any of these scenarios, having one or more architects (solution architect and/or technical architect) available to support the development team in coming up with alternative or new solutions *ensures velocity and alignment with target architecture*.

  - Depending on the scale of your Salesforce project (or program), architects may be partially or fully allocated to your project but usually need not be fully allocated to a single agile team.

- **The business analyst (BA)**:

  - In the development phase, the BA partners with your PO to refine user stories and get them **ready for development**. The definition of ready typically includes user story persona, need, purpose, additional context and supporting documents, and, critically, acceptance criteria. Depending on your degree of upfront architecture and level of user story solution design, your process may involve the BA to assign the user story to a solution or technical architect to provide solution design before the user story meets the definition of ready.

  - The role of BA is sometimes taken up by a functional consultant in the development team as they – stereotypically – are more "business-oriented" and, therefore, should be more suited to interact with the PO and other business stakeholders to support them in the creation of user stories.

> **Important note**
> Many functional consultants can also act as great BAs. But it is not always the case. Work with your implementation partner to make sure your agile team has solid business analysis skills in the team as the BA plays a key role in the success of your Salesforce project.

- **The quality assurance (QA) specialist (tester)**:

  - QA specialists are responsible for *testing user stories against the acceptance criteria*. They do this by writing positive and negative test cases based on the acceptance criteria of a user story and executing the test case to evaluate whether they pass or fail.

    - The QA specialist (or test manager in larger projects) will design the overall test strategy and testing regime of your Salesforce project.

- Experienced QA specialists are also experts in *trying to break solutions* and processes, which is exactly what you want them to do in the development phase. You want to understand risks and identify **defects** (bugs) in your Salesforce solution early in the development phase so they can be fixed now rather than in **user acceptance testing** (UAT) or, worse yet, in production, thus impacting productivity and risking adoption and, ultimately, the value of your investment in Salesforce.

- **The change manager**:

  - The change manager will work with your agile team, architects, business stakeholders, and project sponsors to co-create a *context-specific change management strategy and plan*. We'll cover specifics of change management planning in *Chapter 10, Communicating, Training, and Supporting to Drive Adoption*.

- **The project manager (PM)**:

  - The PM is responsible for the overall delivery of your Salesforce project. This includes ensuring adequate resources are timely available in the project, managing issues, mitigating risks, reporting on project progress, and managing the budget for your Salesforce project.

  - When you're working with an implementation partner, it is beneficial to have an internal client-side PM to partner with the PM from your implementation partner.

  - Depending on the scale and number of agile teams in your Salesforce project:

    - The scrum master and PM roles may be taken up by one person from your implementation partner.

    - An **engagement manager** from your implementation partner may be part of the project organization to liaise with your internal project leadership and sponsor to plan beyond the current implementation project, act as an escalation point for risks and issues, and be responsible for managing the commercial relationship and contract.

Next, let's discuss how you can plan development.

Planning your build phase

In a pure agile delivery methodology, the team plans their work continuously – either in sprint planning (Scrum) or by simply starting new work when new user stories meet the **definition of ready** and there is the capacity for it to be picked up (Kanban).

However, having chosen a form of hybrid agile delivery methodology, you'll likely want to have some indication of the estimated duration of the build phase – both for budgeting and roll-out planning purposes.

In this section, we'll discuss the advantages and disadvantages of various methods of **estimating development effort** and how you can plan development.

Let's look at two methods of estimating effort:

- **Estimating development effort in time format**:

  - In the previous chapter, you broke down the overall scope into user stories and provided solutions for each of them. If you didn't, go back and review this process.

  - To know the total number of estimated development hours, you would need to have each user story estimated (time: hours) *by development type* (configuration, data, and customization). Inherently, for an estimate to be somewhat accurate and hence useful for planning purposes, each user story needs to have a solution design done beforehand.

  - Once all stories are estimated, you summarize the estimates by epic and then summarize all epics by theme/domain to know the total.

  - It is important to note that, at this point, the estimate is simply an estimate. It is not until the story or sprint is finished that you will know exactly how long it took.

Important note

When summarizing the estimate effort - and thus cost - in the development phase, it sometimes exceeds the original budget for your Salesforce project. This is typically not because the implementation partner is trying to squeeze you. It is usually because the consultants ask questions and uncover requirements, rules, exceptions, edge cases, and enablers, which weren't apparent to you before the scope was detailed.

- **Story points**:

  - Story points are radically different from estimating in a time format. A user story's story points *describe a mix of total effort, complexity, risks, and uncertainties* expected for the agile team to refine, design, develop, test, and deploy and release a user story. As you can see, the story point concept extends far beyond the hour estimate model, which focuses on the time for the development of the user story.

  - A key aspect of story points is that they are *relative to an agile team's history of working together in the team*. This also means that assigning story points to all user stories upfront isn't a meaningful exercise.

  - Story points are typically assigned to user stories in sprint planning. We'll return to this later in the *Maximizing value from agile ceremonies* section.

  - Using story points is suitable for projects where less effort has been made up front to do user-story-level solution design.

After estimating your development effort, let's move on to creating your development plan.

Drafting your tentative development plan

Assuming you have a budget and a deadline for your project, you need a plan for your development to know how you will respect them.

Once you have each user story estimated, and thus each epic, you need to consider your available development capacity. This is the amount of time your functional consultants and developers have available for development tasks. You need to be mindful of the development estimates by development type. Next, you need to factor in time for agile ceremonies and possible backlog refinement, for which – although they are important sessions – the time is not available for development. The difference is referred to as **sprint efficiency** and is typically between 50% and 70% depending on your chosen delivery methodology and hence, the degree of upfront user story definition, detailing and providing solutions.

Tip

As a general rule of thumb, you shouldn't fill up more than ~85% of your agile development team's net development sprint capacity (after applying sprint efficiency). The reason being when you progress through sprints – and see your solution come to life – you will be inspired to do more things or to modify a process that made sense in workshops but isn't ideal for users after all.

If the development capacity (resources) available in your project doesn't match your total estimated effort or timeline, you have three options:

- *Prioritize* what user stories are in the scope of the initial release of your Salesforce solution.

- Work with your implementation partner to *find alternatives* – more manual solution options – for some user stories.

- *Increase the budget* by allocating additional development resources to deliver the project according to your initial timeline. There is a limit to how many development resources you can add and still gain added capacity. It will depend on the nature, scope (number of domains), and the sophistication of your development setup. We will cover development model and process in the next section. Alternatively, you can consider prolonging the build phase to be able to deliver the scope.

Having determined your scope and available development capacity, you are ready to create your tentative development plan by assigning epics to sprints accordingly by taking dependencies into account.

Let's have a look at an example from a scenario company, PME:

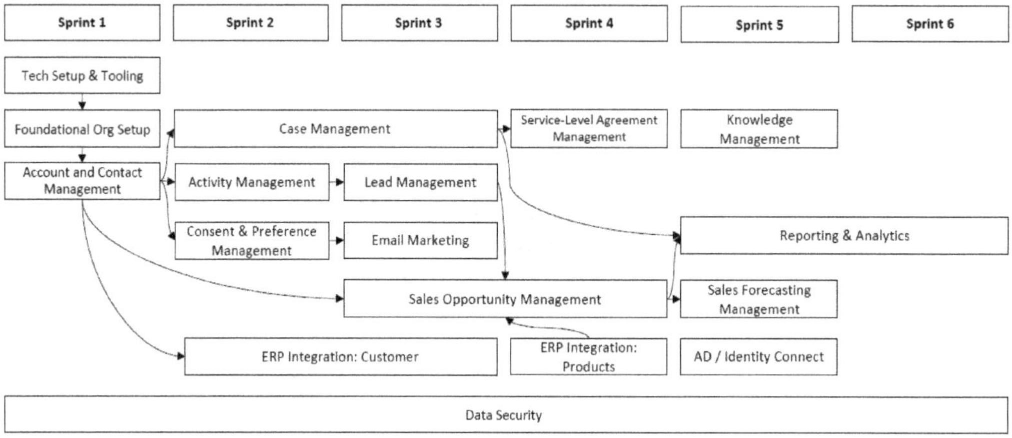

| Sprint 1 | Sprint 2 | Sprint 3 | Sprint 4 | Sprint 5 | Sprint 6 |

Tech Setup & Tooling

Foundational Org Setup — Case Management — Service-Level Agreement Management — Knowledge Management

Account and Contact Management — Activity Management — Lead Management

Consent & Preference Management — Email Marketing — Reporting & Analytics

Sales Opportunity Management — Sales Forecasting Management

ERP Integration: Customer — ERP Integration: Products — AD / Identity Connect

Data Security

Figure 7.1 – PME's tentative development plan

PME's development plan has dependencies between its functional epics and technical enablers. In addition to the functional scope, the technical foundation to start development needs to be established before actual development can commence. This includes setting up technical tooling for the development pipeline, creating development environments (sandboxes), and creating users and test users for the development team and the PO.

After you've created your development plan, let's get your development life cycle and deployment process set up.

Setting up your development model and environments

The **development model** and environments are part of your **deployment model**. Deployment means moving the technical components of your solution from one environment to another and eventually to production. You use different environments – **sandboxes** – to perform different types of testing in your development life cycle. The purpose of this is to make sure your solution works end to end as intended when it reaches production.

Let's begin by looking under the hood of Salesforce.

Understanding the components of your Salesforce solution

The products Salesforce offers, such as Sales Cloud and Service Cloud, are cloud-based **software-as-a-service (SaaS)** products. That means your Salesforce org – depending on the licenses you have bought – comes with standard objects and standard fields **out of the box (OOTB)**.

Salesforce is also a **platform as a service (PaaS)**, which means you can extend, enhance, configure, and customize the OOTB components within Salesforce to fit your business needs and processes. This includes creating custom objects and fields and automation, configuring and customizing the **user interface (UI)**. You can also use the Salesforce platform to build custom apps and integrations. The configuration and customization capabilities that allow you to quickly configure and customize Salesforce to fit your business needs are one of the things that makes Salesforce a powerful platform.

To better understand the configuration and customization of your solution, we need to understand the underlying architecture of the core Salesforce platform, which is based on a **model-view-controller (MVC)** architecture.

Let's have a deeper look at what MVC consists of by seeing some examples:

	Model (data structure)	**View** (UI)	**Controller** (logic)
OOTB components	Standard objects, standard fields, and standard picklist values	Standard apps, default page layouts, and Lightning record pages	Standard duplicate/ matching rules and lead conversion
Configuration components (no-code)	Custom objects and custom fields (including relationship fields), custom metadata types, and custom settings	Custom apps, page layouts, Lightning record pages, and screen flows	Business logic: flows, process builder (retiring), and workflow rules (retiring) Integration: connectors, flow with outbound messages and enhanced External Services
Customization components (low-code/pro-code)		LWCs, **Visualforce (VF)** pages, Aura components, Canvas Apps, and iFrame	Business logic: Apex triggers, Apex classes, and Salesforce functions Integration: Platform events, enhanced External Services, web services, OData, and standard APIs

AppExchange components	Can consist of a combination of the configuration and customization MVC components

Table 7.1 – The MVC components of a Salesforce solution

Having understood the technical components that make up your Salesforce solution, let's see what options are available for you to develop and deploy your solution.

Introducing development models in Salesforce

A development model is a method for migrating your solution through environments, and there are different models available to you, each with its own considerations.

The following table compares the available Salesforce development models:

Development model	Source of truth	Release artifact and deployment method	Level of sophistication, flexibility, and scalability
Manual deployment	Salesforce production org	Manually recreating configuration	Extremely low
Change Set Development Model	Salesforce production org	*Metadata* deployed with change sets	Low
Org development model	Salesforce production org or source control repository (if used)	A .zip file with metadata deployed with the Salesforce CLI or another DevOps tool (see the next tip box)	Medium
Modular package-based development model	Source control repository	**Unlocked package** containing a logical set of components to support functionality for a group of users deployed by installing the latest *version* of the unlocked package	High

Table 7.2 – Salesforce development model comparison

You should have made a conscious decision together with your implementation partner about which development model is to be used for your Salesforce project.

> **Tip**
>
> Several **commercial off-the-shelf (COTS)** DevOps or CI/CD tools are available to do most of the heavy lifting and provide helpful insights into the most common deployment errors. Some of the most popular products include *Gearset, Copado, Salto*, and *AutoRABIT*. Perform your own research and evaluate whether a DevOps tool is beneficial for your Salesforce program at this stage of its life cycle.

The decision about the development model and DevOps tool needs to consider who will be managing your Salesforce org and continuous improvement *after* the initial release of your solution. If you plan to manage your Salesforce org yourself, the chosen development model should align with the available resources' competencies – or planned training and enablement.

Let's look at the environments you should consider for development and testing in Salesforce.

Understanding Salesforce sandboxes

When creating a Salesforce sandbox, you are creating a copy of your Salesforce production environment. You can also clone a sandbox and create a new sandbox that way.

If your Salesforce program/project is following either the org-based development model or the package-based development model, you should be using a version control repository as the source of truth from which you are migrating your metadata to new sandboxes.

You need to use sandboxes to keep your development and testing environments separate from each other and separate from production. You should never develop in production because the implication of a broken feature can impact business operations, corrupt your data, and risk security compliance.

You need separate environments for development, and for different types of testing – determining which sandboxes to use is what is referred to as your **environment strategy**.

Let's take a look at the sandbox setup and deployment flow of our scenario company, PME, which has multiple development streams with separate teams working on its Salesforce solution:

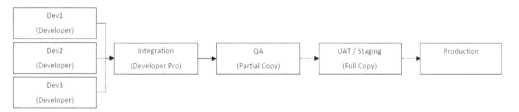

Figure 7.2 – PME's sandbox setup

Let's look at the key differences between the available Salesforce sandbox types in the following table:

Sandbox	Full Sandbox	Partial Copy Sandbox	Developer Pro Sandbox	Developer Sandbox
Appropriate use	System integration testing (SIT), UAT, batch data, performance or load testing, staging, and training	QA, SIT, UAT, and training	Development and QA	Development and QA
Included at refresh	Metadata and all data	Metadata and sample data	Metadata	Metadata
Available storage	Same as prod	Data: 5 GB File: Same as prod	Data: 1 GB File: 1 GB	Data: 200 MB File: 200 MB
Sandbox Templates	Available	Required	Not available	Not available
Refresh interval	29 days	5 days	1 day	1 day
Cost as a percentage of net spend with Salesforce – per additional sandbox	30%	20%	5%	N/A

Table 7.3 – Salesforce sandbox type overview

The number of different types of sandboxes available to you depends on the edition of your Salesforce org. You can purchase additional sandboxes. Refer to the *Sandbox Licenses and Storage Limits by Type* **Salesforce Help** article at `https://help.salesforce.com/s/articleView?id=sf.data_sandbox_environments.htm&type=5`.

If you are developing using the org-based development model or the package-based development model, you may opt to use **scratch orgs**. They are short-lived (1-30 days) development environments that are created based on a configuration file. Working with scratch orgs requires a more advanced skillset compared to working with traditional Salesforce sandboxes.

> **Tip**
>
> You can learn more about Salesforce development models and environments by following the Trailmix *Architect Journey: Development Lifecycle and Deployment* page at `https://trailhead.salesforce.com/users/strailhead/trailmixes/architect-dev-lifecycle-and-deployment`.

You can learn more about scratch orgs in the *Salesforce DX Developer Guide* documentation at `https://developer.salesforce.com/docs/atlas.en-us.sfdx_dev.meta/sfdx_dev/sfdx_dev_develop.htm`.

After determining your development model and setting up your sandboxes, let's turn to your development process.

Establishing your development process

One of the key components of the development process is the **sprint**. A sprint is a term used in **Scrum** to describe a predefined duration in which the aim is to develop working software. We have already mentioned that a sprint's duration is not set in stone – it can be decided that it will last anywhere from 1-4 weeks. Let's look at what you should consider when determining your sprint duration:

- Will user stories need refinement during the sprints?

  - *If yes*, this will take time away from development. Having a short (1-2 weeks) sprint duration will impact the expected sprint efficiency beyond what is sensible, as the overhead for refinement and agile ceremonies is greater than the available time for development. If you need continuous backlog refinement sessions, increase your sprint duration.

  - *If no*, meaning you have already refined and provided solutions to your user stories upfront, then you may opt for a shorter sprint duration, though a 1 week sprint duration is not recommended when building an initial release.

> **Important note**
>
> It's a common misconception that agile requires less discipline compared to waterfall. The opposite is true: *running a pure agile setup requires enormous discipline*, well-defined processes, and commonly agreed-upon definitions of ready and done. In waterfall – in any one phase of a project – there is only one overall task at hand, for example, designing a solution. In pure agile, on the other hand, your team needs to go through the entire development life cycle (or most of it) from user story creation, design, build, deployment, testing, and validation. With many of the steps being performed by different team members, a high level of discipline is required to ensure a smooth handover and clear processes. The benefit of running a pure agile setup is also directly correlated to the combined experience and agile maturity and discipline of the team.

Let's dive into the development process in more detail.

Maximizing value from agile ceremonies

In order to make the most of your agile ceremonies, you first need to have a solid understanding of the nature, purpose, and value of them.

Let's have a look at agile ceremonies:

- **Sprint planning**: The goal of sprint planning is to have a *committed sprint backlog* of user stories and a clearly articulated sprint goal. Before sprint planning, the PO should order the backlog – this is sometimes done together with the BA in a **backlog refinement session**. A typical agenda for sprint planning goes as follows:

  - Review team capacity for the sprint:

    - If you have estimated your user stories using the time format, consider the available resources in the development team and account for sprint efficiency.

    - If you're using story points, also consider available resources and adjust the target story point for the sprint based on prior sprint velocity. For the first sprint, this is tricky as there is no history: rely on the development team's assessment as you go through the user stories.

  - Next, the PO presents user stories one at a time, and the development team asks clarifying questions.

> **Tip**
> You may or may not require user stories to meet your definition of ready for them to be eligible for sprint planning – and to be able to be pulled into the sprint. If you allow this, make sure the PO, BA, and architects allocate time at the beginning of the sprint to get them ready for development.

  - The development team pulls user stories into the sprint backlog until the development team's capacity is met.

    - If you have estimated your user stories using the time format, capacity is met when the cumulative sum of the user stories pulled into the sprint reaches the team's capacity for the sprint.

    - If you're using the story point method, team members provide an estimate for the user story using **story point pokering**, whereby a number in the Fibonacci sequence is used. If the estimates are close, the team can settle easily on a story point for the story and move on to the next user story. If there's a great disparity between estimates, the team members with high and low estimates voice their reasoning for their estimates. This is a great way to make sure team members share an understanding of the user story and perhaps uncover complexity that wasn't obvious to all members. The team settles on a story point for the story and moves on.

> **Tip**
>
> For virtual story point pokering, `https://www.scrumpoker-online.org` is a great online tool.

- When the team capacity in story points is met, the stories that have been pulled into the sprint are committed.

- **Daily stand-up**: This is a short (~15 minutes) session, typically every morning, where each agile team member answers three simple questions: what did I do yesterday to progress towards the sprint goal, what am I planning to do today to progress towards the sprint goal, and what blockers are preventing me?

  - The daily stand-up is *not a project status meeting*, nor is it a meeting for resolving blocking issues. Identified actions should be executed offline after the stand-up.

- **Sprint review (demo)**: A sprint review is a session where your agile team shows a wider group of stakeholders what was completed in the sprint. The sprint review serves two main purposes:

  - To show your solution to select end users (change agents), thereby embedding change management throughout the development phase.

  - To get feedback on the solution that can be assessed and potentially incorporated into later sprints. This creates a sense of ownership for the participants in the sprint review – laying a solid foundation for them to act as ambassadors for your solutions when it's time to roll out.

> **Tip**
>
> Ideally, the PO from the agile team drives the sprint review, demos what was developed, and responds to feedback from the participants. This will ensure the solution ownership is clear, and also that it is gradually internalized by your PO and your company.
>
> While it may be the responsibility of your implementation partner to consolidate the feedback from participants, your PO is ultimately responsible for determining the priority of each feedback item and communicating decisions back to the stakeholders.

- **Sprint retrospective**: In this session, the agile team members, PO, scrum master, and the development team evaluate how the past sprint went and what improvements they can make to optimize the development process in the spirit of continuous improvement.

  - It is facilitated by the scrum master, who is also responsible for making sure actions agreed upon in the sprint retrospective are executed.

Let's look at the flow of user stories before and during a sprint.

Managing your Kanban board

How your agile team manages the flow of user stories has a great impact on the sprint velocity. The members of your agile team should know your development process and, critically, how handover is done.

Let's take a look at how a Kanban board could be set up (the specific configuration for your Salesforce project will vary based on your chosen delivery methodology):

Figure 7.3 – Kanban board columns example

A new user story will start in the **Product Backlog** column. Your PO will work on the user story together with the business analyst to progress the user story to meet the definition of ready. This, again, will depend on your chosen delivery methodology but should meet the following criteria:

- The user story is complete with persona, need, and purpose.

- Additional context and supporting documents are provided.

- Acceptance criteria are defined.

- In some development processes, the user story must also have a solution design provided and approved by a design authority. We'll cover this in more detail in the *Governing your Salesforce project in the development phase* section.

When a user story meets your definition of ready, your PO moves it to the **Ready** column. The story is ready for sprint planning.

User stories are pulled into the **Sprint Backlog** column during sprint planning. As the sprint begins, the development team will pick user stories at the top and move them to the **Development** column. Each development team *member* should only have one user story in development at a time.

> **Tip**
> While practical Salesforce configuration and development are not in the scope of this book, there are several other great books that cover those topics, including *Low-Code Application Development with Salesforce* by Enrico Murro (https://www.amazon.co.uk/Hands-Low-Code-Application-Development-Salesforce/dp/1800209770).

If a development team member is blocked from completing the development of a user story, they should move it to the **Blocked** column and notify the scrum master by writing a comment on the user story. The scrum master will prioritize resolving the blocking issue.

> **Tip**
>
> In order to minimize the risk of user stories going stale or becoming outdated, and reduce the waiting time because of dependencies, you should aim to have user stories in development (to the right of the Sprint Backlog Kanban column) for the shortest time possible. The way to ensure this is for all agile team members to work on user stories assigned to them *starting from the right side* of the Kanban board. For example, your PO should *prioritize* validating user stories and moving them to the **Done** column *before* focusing on getting user stories ready for development. Similarly, a developer should *prioritize* addressing defects identified by a QA specialist *before* starting development of a new user story.

When a user story is developed and deployed to a QA sandbox, the development team member moves the user story to the **Testing** column and writes a comment on the user story, letting the QA specialist know it is ready for testing, along with any specific instructions. In some cases, the developer of a user story is required to provide a test script.

Following a successful test pass, the QA specialist moves the user story to the **Validating** column for the PO to validate the user story against acceptance criteria. If the user story did not pass testing, the QA specialist will create a defect issue describing what failed and then move the user story back to Development and assign it back to the developer to fix the issue.

When a user story is validated by the PO, the PO will move the user story to Done.

> **Important note**
>
> This chapter describes the user story flow in the context of building an initial release of your Salesforce solution – you are not yet deploying and releasing it to production for users to benefit from. We'll cover that process in *Chapter 9, Deploying Your Release and Migrating Data to Production*.
>
> After your initial release is live in production, you will enter a phase of continuous improvement with continuous development and deployment of user stories. In that setup, your definition of done will include more steps than described in this chapter. In *Chapter 12, Evolving Your Salesforce Org and DevOps Capabilities*, we'll discuss this in more detail.

Next, let's look at how to govern your Salesforce project through the development phase.

Governing your Salesforce project in the development phase

There are many things happening in the development phase of your Salesforce project, and it can be overwhelming to comprehend it all and get an overview. Your responsibility in the building and testing of your initial release is simple:

> *To create an environment, a structure, and processes to maximize the high-quality value output of your development team.*

At a high level, you want to ensure your agile development team is:

- **Building the right thing**: This means ensuring your Salesforce solution will:

    - Be well received and used by the end users

    - Meet your objectives and financial targets set out in your business case

- **Efficient**: This involves making sure your agile development team is building your solution in the most efficient manner possible:

    - **Value**: Output x quality

    - **Cost-efficient**

In the following sections, we'll break down what this means in practice.

Salesforce project governance

There are effectively two key aspects of governance you need to manage in your Salesforce project:

- Salesforce project *delivery governance*

- Salesforce *platform governance*

To deliver your Salesforce project in accordance with these aspects, you should consider putting in place the following governance mechanism:

- **Design authority (DA)**: In this forum, proposed solution designs from the development team members or architects are continuously reviewed to ensure the solutions introduced to your Salesforce org align with the target solution architecture, abide by your defined development standards, and follow general best practices for declarative and programmatic development and architecture.

- Participants in the DA:

    - There are many configurations of DAs, but the key thing to note is that it is not purely a technical meeting. It is critical that your PO(s) participates in presenting the background of the user story (or new epic).

    - A technical architect and solution architect should participate in reviewing proposed solution designs and provide feedback and suggestions for improvements to the design to the development team.

- The focus of the DA:

    - The DA should focus – from a risk-based perspective – on proposed solutions, which include the following solution components:

        - Code

        - Data model changes

        - Security and data sharing

        - Complex automation

        - Integration

    - The DA should be mindful of the degree of customization of the aggregated proposed solutions.

- **Change control board**: As you progress through the sprints in the development phase, you will become more familiar with your Salesforce solution. Your PO will likely be inspired and will want to include additional features beyond the originally planned scope. To manage this process (common in fixed price/fixed scope projects), you can establish a change control board to address and manage change requests in the development phase.

    - There should be a threshold for what level of change your change control board should be involved in:

        - **Small change**: Empower your agile team to manage these

        - **Medium change**: Possibly address these to your change control board, depending on their impact on the project timeline and risk

        - **Large change**: These inherently impact the project timeline and must be presented to your change control board

    - Recommended participants: PO, project sponsor, PM, and Salesforce architect

- **Steering committee**: This is the highest escalation point for your project. The steering committee is accountable for the project and executes this by providing guidance and support to the PMs and architects. It meets on a regular cadence (bi-weekly or monthly) to do the following:

  - Track the development progress and status against the plan

  - Track the project's spend versus budget

  - Address issues and mitigate risks

> **Important note**
>
> Shield the development team members from this project meeting.

> **Important note**
>
> The most optimal delivery governance *configuration* for your Salesforce project depends on the nature and scope of your Salesforce project and the number of agile teams. If you have multiple teams or streams, you should consider following one of the more advanced **Scaled Agile Framework (SAFe)** configurations.

Let's dive deeper into testing your Salesforce solution.

Governing the quality of your Salesforce solution

Your implementation partner may be responsible for the quality of the solutions they deliver, but your company is ultimately accountable for the solutions you ask your company's users, customers, and partners to use and interact with. Therefore, it is important that you are familiar with the key methods of testing and quality assurance.

Let's go over the two types of testing in Salesforce you need to understand:

- **User story testing (functional testing)**: This is a commonly used manual test method whereby a single functionality (or user story) is tested to ensure it works as expected. The tester should ensure both positive and negative test cases prove the solution works. We described the details previously in this chapter, in the *Managing your Kanban board* section – as it should be done for each bit of configured or developed functionality.

- **Unit testing**: This is a test type enforced by Salesforce for each deployment (with code) to have a test coverage of at least 75%. But it can – and should – also be used for testing business-critical processes, such as lead record creation, contact record creation, and lead conversion. The upfront effort for creating test classes for configuration functionality greatly outweighs the pain of having to fix critical process failures in UAT or production.

> **Important note**
>
> There are many other types of testing that you and your implementation partner can and should perform, including SIT, **performance testing**, and **load testing**. They are, however, beyond the scope of this book. If you want to dive deeper into the fascinating world of software testing, **ISTQB** and **TMAP** are two of the most widely used test frameworks. For larger Salesforce projects, it may be worthwhile for your organization to invest in setting up test automation. Engage with a Salesforce architect to investigate the potential use of a tool such as **Provar** or **Selenium**.

At the end of your development phase, or right at the beginning of your roll-out phase, you will carry out UAT to ensure a larger group of your intended end users test and accept your Salesforce solution before releasing it to production. We'll cover this in *Chapter 9, Deploying Your Release and Migrating Data to Production*.

> **Tip**
>
> As you progress through the phases of your Salesforce project, you are likely to encounter resistance and receive feedback. While it's important to *take feedback seriously – potentially making necessary changes –* and try to understand the underlying causes of the resistance, it is critical that you also *stand firm in your conviction* of the vision for your Salesforce project and trust in your efforts and analysis in the pre-development phase of your project.

Let's look at different ways of tracking the progress of your Salesforce solution development.

Measuring progress in the development phase

Earlier, you estimated the development effort of the scope of your Salesforce solution. To understand how you are progressing, use the following key reports and KPIs:

- **Burndown chart**: This displays story points or hours remaining of a sprint's scope.

- **Version report**: This predicts the date for completion of the remaining user stories in the product backlog. For the version report to be meaningful for your Salesforce project, you need to have all user stories estimated (with either story points or hours). Tools to support the agile development process provide a best- and worst-case scenario based on your team's delivery consistency.

- **Sprint velocity**: Depending on your estimation method, you have these ways to measure sprint velocity:

  - The sum of the story points completed in the sprint.

  - The sum of the stories' estimated development hours completed in the sprint.

- **Added scope**: While added scope may be valuable, it decreases efficiency and adds risk in terms of your project being deliverable on time. There are two levels to added scope you need to be aware of and manage:

    - **New scope**: To manage scope creep and increase development efficiency, it is a good idea to track the number (and estimate) of *new user stories* added to the product backlog.

    - **Late sprint scope**: User stories added to the sprint backlog *after sprint planning* were not discussed in sprint planning, and, as such, the development team members were not able to assess and commit to delivering them. Worse yet, the development team will need to spend time understanding what the user stories are about and how to solve them.

> **Important note**
> While it is tempting to try to get the development team member to "work faster or smarter" to deliver more in a shorter period of time, it is almost always a more effective strategy to look beyond the development team and *improve areas such as organizational alignment, empowerment of your PO, and removal of dependencies between teams.*

Let's move on to talk about your data.

Preparing your data migration plan

As part of your effort and analysis in the pre-development phase (covered in *Chapter 2, Defining the Nature of Your Salesforce Project*) you identified the *systems* that are currently used to support your business capabilities. As you are developing a new solution based on the Salesforce platform, your task now is to identify what data from your **legacy systems** needs to be moved to Salesforce and how to migrate it.

Your data migration strategy should answer the following key questions:

- What *systems* will you be migrating data from?

- What *data* will you be moving?

- How will you *extract* the data from the legacy systems?

- Which Salesforce objects does your data *map* to – where will the data reside?

- How will you *transform* your data to make it consumable by Salesforce?

- How will you *load* the data into Salesforce?

Start by listing all your legacy systems and data (objects) and note the number of records in each object.

> **Important note**
>
> Be sure to identify any potential **large data volume** (**LDV**) objects as they will have an impact when deciding how to extract your data from your legacy system and how to load it to Salesforce. Learn more about LDVs and their implications for your Salesforce org in the article *Best Practices for Deployments with Large Data Volumes* at `https://developer.salesforce.com/docs/atlas.en-us.salesforce_large_data_volumes_bp.meta/salesforce_large_data_volumes_bp/ldv_deployments_introduction.htm`.

If you have decided to *phase your project roll-out* (as part of your analysis in the pre-development phase in *Chapter 3, Determining How to Deliver Your Salesforce Project*), you need to determine the record count for groups of users in your various roll-outs.

You then need to understand how you can extract the data from legacy systems and in what format. Some systems can connect to data **extract transform load** (**ETL**) tools, such as Informatica, which speeds up extraction and transformation.

Next, map which Salesforce objects the data will reside in. This can be a tricky exercise as **data models**, and levels of **data normalization** differ across systems. Engage a Salesforce architect or **data migration specialist** to support you in this.

> **Tip**
>
> Create a **data dictionary** describing all the objects and fields used in your Salesforce solution. Make sure it is aligned and updated continuously by your development team.

You now know the source and the target for the data and the discrepancy in data formats for each attribute or field, and you can determine your method for transformation to make the data ready for Salesforce.

Next, you need to determine how you will load the data into Salesforce. There are many options available – their suitability for your Salesforce project depends on factors such as the size of your datasets, including any LDVs, and how long or short your **data migration window** is. This brings us to the final point: data migration planning.

Some businesses are only open for business during daytime business hours, which makes evenings, nights, and weekends suitable options for your data migration window without interrupting the business. But what if you don't operate in a single time zone or your users potentially also work in the evening and through the night? In those cases, you are faced with some hard decisions and two overall options:

- You can attempt to plan and time your data migration in a finely tuned cumulative manner:

    - This requires advanced orchestration and planning and also requires several **dry-runs** in a staging sandbox to optimize the execution before moving data to production at go live.

- There is another risk to this approach. If the data migration is delayed, the business will end up being disrupted anyway.

• Accept the impact of disruption on the business. An upside to this is you can plan user training to be held while data is being migrated.

> **Tip**
>
> Key tips to maximize your data migration at go live:
>
> • Start data migration planning early in the development phase to define your strategy.
>
> • Prioritize data cleansing. Data quality is a big challenge for many companies. The time you invest in cleaning data and removing duplicates will enhance your user experience and the overall performance of your Salesforce org, leading to better adoption.
>
> • Your data migration strategy should align with your development principles. Make sure there is close collaboration between the people responsible for your data migration strategy and your agile development team.

We'll cover data migration preparedness in *Chapter 9, Deploying Your Release and Migrating Data to Production*.

> **Tip**
>
> To gain a deeper understanding of the intricacies of migrating data to Salesforce and managing it in general, you can read the *Salesforce Data Architecture and Management Data* book written by Ahsan Zafar (CTA) (`https://www.amazon.com/Salesforce-Data-Architecture-Management-effectively/dp/1801073244`).

Let's wrap up this chapter.

Summary

You have completed a lot of activities to progress from solution design to building and testing the user stories of your Salesforce solution. You have also created a plan to migrate data from your legacy system(s) into Salesforce as you roll out your Salesforce solution to users. Having completed these activities, you should evaluate the state of your Salesforce project as you are nearing the end of *Part 2, The Development Phase*.

In the next chapter, we'll discuss typical issues you may encounter in the development phase of your Salesforce project and how to mitigate and overcome them. You'll also be provided with a checklist to evaluate the state of your Salesforce project.

Common Issues to Avoid in the Development Phase

In the previous two chapters, you covered the activities in the development phase of your Salesforce project: detailing the scope of your release, creating the solution design, establishing your team's development process, completing the build and testing of your release, and creating your data migration plan.

In this chapter, we will discuss common issues often faced in the development phase of a Salesforce project. We will also cover strategies to prevent or mitigate such issues from arising altogether.

In *Chapter 5, Common Issues to Avoid in the Pre-Development Phase*, you were introduced to **root cause analysis** (**RCA**), and we'll continue using that method to dissect the common issues faced in the development phase. At the end of this chapter, you will be presented with a set of **checklists** to evaluate the state of your Salesforce project as you wrap up the development phase.

This chapter will cover the following main topics:

- Common issues in the development phase and their root causes

- Evaluating the state of your Salesforce project

Let's go!

Common issues in the development phase and their root causes

Let's get started by looking at some of the issues most commonly faced in the development phase—illustrated with a fishbone diagram:

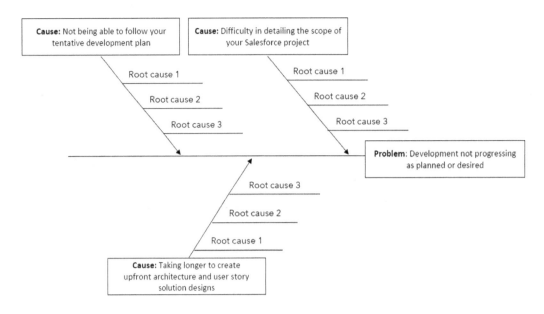

Figure 8.1 – Fishbone diagram of common issues in the development phase

Next, we'll discuss the possible root causes and strategies for the mitigation and prevention *of each* of the common issues.

Not being able to follow your tentative development plan

In *Table 8.1*, you'll find possible root causes for why you may not be able to follow the tentative development plan for your Salesforce project, along with strategies for mitigation if you are already facing that issue:

Possible root cause	Strategies for mitigation and prevention
Changing requirements and priorities of your organization.	Empower and expect your Salesforce **product owner** (**PO**) to step up, take charge, and gain buy-in from your business stakeholders for the prioritization of the initial release.
De-scoping or minimizing technical enablers due to a misguided understanding of a false dichotomy of budget versus value.	Make sure you engage a Salesforce architect to advise which technical enablers are prerequisites for your Salesforce solution.
"Scope creep" or in-flight change requests.	Implement a clear process for managing change requests—establishing a change control board with a regular meeting cadence to assess and address change requests is a good practice for medium-to-large Salesforce projects.

Getting stuck super-optimizing feature epics in initial sprints at the expense of remaining epics.	Encourage your PO to consider the bigger picture and end-to-end solution for all external and internal actors and end users of your Salesforce solution. Make sure the development effort (investment) aligns with the strategic importance of the capabilities in the scope of your Salesforce project.
Poor documentation maintenance and visibility for all project team members throughout the development phase and beyond.	While less exciting to discuss, visible and up-to-date documentation is a catalyst for development effectiveness. Prioritize getting a good structure in place where all project members can access and comment on (and select role updates) requirements and solution documentation.
Insufficient establishment, training, and communication of development principles and guidelines to team members.	Establish your guidelines. Then add a section in the onboarding material/deck for your project with references to the guidelines. Communicate the guidelines and continuously adapt and improve them.
Constraints and bottlenecks impede progress.	Investigate the nature of the constraints—cycle time for user stories in each column of your Kanban board. Your agile development team members are likely well aware of what the root cause is— so, all you need to do is ask, listen, and act swiftly to resolve the bottleneck. This may involve shuffling resources, modifying the development process, or perhaps adding understaffed roles.
The technical tooling and the development environment (sandbox) strategy is hindering your development team. The underlying cause could range from developers configuring the same components and overwriting each other's changes when deploying to a common QA sandbox to the development team deploying metadata using change sets.	Engage with the Salesforce technical architect assigned to your Salesforce project and expect them to identify the specific underlying cause and propose changes to implement to improve the development process and tooling used. We'll discuss this subject in more detail in *Chapter 12, Evolving Your Salesforce Org and DevOps Capabilities*.
Neglecting change management efforts.	Make sure all stakeholder groups are represented and informed about your Salesforce project.

Low stakeholder engagement and poor feedback at sprint reviews (demos).	Consider the following actions: • Invite relevant stakeholders to relevant demos • Group user stories to tell a larger overall story • Speak the business language and use realistic test data—focus on the business value of a user story rather than the technical intricacies and sophistication of the development
The underlying assumptions for the development estimates prove false. The expected sprint efficiency could be lower than expected due to more sessions being required for backlog refinement ahead of sprint planning.	Quickly attempt to understand whether the assumptions are generically wrong, which would mean the estimates are wrong or too low for all user stories, or if only a subset of user stories is affected by wrong assumptions. Adjust your development plan accordingly.
Inadequate development team time allocation and skills: scrum master, business analysts, developers, and QA specialists.	Why is this the last point? These team members are likely the ones most experienced in delivering Salesforce projects. While you may encounter inexperienced Salesforce professionals, odds are the root cause(s) for your Salesforce project lagging behind your tentative development plan are to be found and resolved by looking at points proceeding this one.

Table 8.1 – Possible root causes for not being able to follow your tentative development plan

Let's look to the next common issue in the development phase.

Taking longer to create upfront architecture and user story solution designs

In *Table 8.2*, you'll find possible root causes for why it might be taking longer than expected for you to create upfront architecture and solution designs for user stories in the scope of your Salesforce project:

Possible root cause	Strategies for mitigation and prevention
Insufficient allocation of internal technical resources to interact with your implementation partner's architect(s).	Technical architecture such as integration strategy, integration pattern determination, data mastering, and **identity and access management (IAM)** all need to be aligned with your enterprise's practices and capabilities. Make sure sufficient resources from your IT organization are allocated to your Salesforce project.

Technical assumptions you made in the pre-development phase prove to be invalid.	As we discussed in the pre-development phase of this book, you make a lot of assumptions before starting the development phase of your Salesforce project. If you find an assumption to be wrong, assess the impact and adjust your design and/or plan accordingly.
The quality of user stories and their acceptance criteria are unfit to be solutioned against.	Enlist the PO of your Salesforce project in a crash course on the concept of user stories—ideally, the business analyst can host this session. Additionally, your PO can review the Salesforce Trailhead module on the subject: `https://trailhead.salesforce.com/content/learn/modules/user-story-creation/construct-a-user-story`.
Insufficient resources assigned with skills and experience in the intricacies of Salesforce architecture.	Salesforce is a vast platform spanning traditional CRM, analytics, integration, PaaS capabilities, and collaboration tools. You rarely encounter an architect who masters all aspects and products. Make sure you have adequate Salesforce architect resources in your Salesforce project.
Difficulties in determining your development model, sandbox and development life cycle, and deployment strategy.	Engage with and expect the technical architect of your implementation partner to drive the analysis and recommendation of these topics. Your senior IT leadership stakeholders should be closely involved in the discussions and buy into the decisions made as it impacts your license costs and operating model.
The scope of your Salesforce project is too large (8+ sprints of 2 weeks) for your project team to comprehend and effectively design and estimate upfront.	Consider phasing your Salesforce project over two or more larger releases. Alternatively, consider only creating a user story-level solution design for the initial sprint and allocate resources for a *separate, concurrent design stream.* If you choose to implement the latter approach with a separate design stream, it is critical that the solution designs for user stories for a given sprint must be completed and reviewed by a design authority prior to the start of that sprint. If not, there is the risk of some of the user stories not being completed during the sprint, which in turn requires more sprints and an extension of the development plan.

Table 8.2 – Possible root causes for taking longer to create upfront architecture and user story solution designs

Let's jump to the next common issue in the development phase.

Difficulty in detailing the scope of your Salesforce project

If you are yet to detail the scope of your Salesforce project, you can read along to know what to be mindful of—and then recap *Chapter 6, Detailing the Scope and Design of Your Initial Release*.

In the following table, you'll find possible root causes for why you might be facing difficulties in detailing the scope of your Salesforce project, along with strategies for mitigation if you are already facing that issue:

Possible root cause	Strategies for mitigation and prevention
The vision and initial high-level scope for your Salesforce project are unclear or not determined before starting the development phase.	Identify and align the vision and high-level scope with the sponsor of your Salesforce project.
Team members are not properly onboarded and introduced to your Salesforce project.	Create an onboarding deck consisting of key project information to enable new project team members to understand your Salesforce project. Include at least the vision and nature of your Salesforce project, project team members' roles and responsibilities, delivery methodology, development guidelines, tools and processes to be used and followed for development and deployment, and available project documentation (scope, requirements, user stories, and design).
Inadequate internal resources are assigned to your Salesforce project. Oftentimes, an organization embarks on the journey to implement Salesforce—while also wanting to try out an agile or hybrid agile delivery methodology—and it becomes overwhelmed with the level of required involvement of internal resources throughout the project. Specifically, it may be your PO lacks experience, skills, time allocation, or empowerment/mandate to take on the big responsibility of the role.	There is no easy way out of this. To have a Salesforce solution that your end users will want to use and that improves your customer's experience of interacting with your organization, you need to invest the time of your key internal resources to provide input to co-create the solution. If these people have never engaged in agile or hybrid agile projects, make sure they are properly upskilled/trained before being allocated to the project, and consider engaging an agile coach to be part of your project.

Inadequate business analysis resources in your project team. Oftentimes, experienced functional consultants are asked to act as business analysts to run workshops, illicit requirements, and challenge the organization's assumptions and old way of working. However, solid business analyst skills are distinct from declarative development on the Salesforce platform. While some functional consultants are great business analysts, not all are.	To remedy this, work with your implementation partner to ensure you have sufficient business analysis skills to support and coach your PO with creation of user stories and solid acceptance criteria.

Table 8.3 – Possible root causes for difficulties in detailing the scope of your Salesforce project

Next, let's evaluate your Salesforce project in the development phase.

Evaluating the state of your Salesforce project

We have covered a lot of topics and activities in the development phase of a Salesforce project. If your project gets stuck, or you are not pleased with the progress, the following subsection will support you in evaluating the state of your Salesforce project. For each item you cannot check, go back to the corresponding chapters and complete the activities to proceed with your Salesforce project.

Checklist for your Salesforce project in the development phase

In the development phase, you should have gone through the following actions. When detailing the scope of your release, you should:

- Make sure the required internal resources are allocated to your Salesforce project
- Involve end users in the co-creation of business processes to be supported by your Salesforce solution
- Make sure the PO for your Salesforce project is suitably qualified and empowered to detail user stories and provide acceptance criteria on behalf of the business stakeholders

Next, you should complete these activities to create the architecture and design of your solution:

- Determine how much **architectural runway** you want to create before starting development
- Have the artifacts described in *Chapter 6, Detailing the Scope and Design of Your Initial Release*, created and understood by your internal Salesforce project team members
- Understand how customized your Salesforce solution will be and the implication on maintainability

Next, you should complete these activities to establish your team's development process to complete the build and testing of your release:

- Made sure everyone in your Salesforce project understands each other's roles and responsibilities
- Understood the development effort of your Salesforce solution (assuming you went with estimating time rather than story points)
- Created a tentative development plan allocating the epics in the scope of your Salesforce project to sprints based on estimated development effort
- Determined which agile ceremonies you will use and how you will maximize value from them
- Defined a **definition of ready** (**DoR**) and a **definition of done** (**DoD**) for user stories
- Determined the practical use of your Kanban board

Next, you should consider putting in place the following mechanisms to govern the development of your Salesforce solution:

- A **design authority** to ensure proposed solutions adhere to your target solution architecture and development guidelines
- A **change control board** to have a forum to discuss and approve or reject proposed changes or new requested scope
- A steering committee to guide and govern the delivery of your Salesforce project
- Made sure an adequate testing regime is in place to secure a high quality of your Salesforce solution
- Understood the key methods to track and understand progress in the development phase
- Consciously considered using a formal framework (such as **Scaled Agile Framework—SAFe**) if your Salesforce project has multiple tracks or streams
- Completed the steps to create your data migration strategy and plan

Let's wrap up this chapter and the development phase.

Summary

In this chapter, you have become familiar with the common issues often faced in the development phase and learned how to mitigate or prevent them from arising. You have also assessed the state of your Salesforce project in the development phase by going through checklists.

With your Salesforce project in great shape, you are ready to move on to the next step in your Salesforce journey, the roll-out phase, starting with *Chapter 9, Deploying Your Release and Migrating Data to Production.*

Part 3: The Roll-Out Phase

This part will take you through the critical activities in the roll-out phase of your Salesforce project and discuss common issues so you can avoid or mitigate them.

This part has the following chapters:

- *Chapter 9, Deploying Your Release and Migrating Data to Production*
- *Chapter 10, Communicating, Training, and Supporting to Drive Adoption*
- *Chapter 11, Common Issues to Avoid in the Roll-Out Phase*

Deploying Your Release and Migrating Data to Production

Following the building and testing of your Salesforce solution in the previous chapters, in this first chapter of *Part 3, The Roll-Out Phase*, of your Salesforce project, you will be guided through the key activities and milestones in the **roll-out phase**. Next, we will dive into some of the key considerations and practicalities of **deploying** your Salesforce solution, as well as how to make the most of **user acceptance testing** (**UAT**). Next, we will evaluate your preparedness for migrating your data to production and provide you with a final go-live checklist to verify you'll be ready to **go live** after training your users.

By the end of the chapter, you will understand the key milestones, outcomes, and activities in the roll-out phase of your Salesforce project. You will gain insights and practical advice for performing **system integration testing** (**SIT**) and driving UAT. Finally, you will understand that planning and performing dry runs is key to successful **data migration** and what to check before signing off for users to go live following training.

In this chapter, we'll cover the following main topics:

- Overview of the roll-out phase of your Salesforce project
- Deploying your Salesforce solution through environments
- Driving UAT of your solution
- Checking your data migration preparedness and signing it off to go live

Let's go!

Overview of the roll-out phase of your Salesforce project

In this phase of your Salesforce project, the focus is on getting your initial solution and legacy data to production and getting users set up for success.

Let's take a look at what you will be doing in this phase.

Key activities and milestones of the roll-out phase

The roll-out phase of your Salesforce project begins when you have completed building the first version – the **initial release** – of your Salesforce solution.

Historically, many organizations – and their implementation partners – have defined a successful roll-out to be when the solution is deployed to production, data is migrated, users have gone through training, logged in for the first time, and the number of new daily production support tickets drops below a predetermined threshold. Sounds familiar?

But to maximize value from your investment in Salesforce, you must go beyond that to define success in the following way:

> *A successful roll-out phase should be defined as the point when your organization, customers/partners, and users are using and gaining value from your Salesforce solution, and it delivers the target business outcomes you laid out in Chapter 4, Securing Funding and Engaging with Salesforce and Implementation Partners.*

As such, to plan for a successful roll-out, we can summarize the key activities and milestones here (the highlighted activities are often overlooked in planning):

- Deploying your Salesforce solution through environments
- Driving UAT of your solution
- Migrating your data to Salesforce
- Signing off to go live
- *Planning, communicating, and managing change*
- Training your users
- Providing hypercare and managing production support
- *Monitoring and driving adoption of your Salesforce solution*

Having established *what* the overall activities and milestones consist of, let's look at *how long* your roll-out phase may take.

Duration of the roll-out phase

Depending on the following factors, the roll-out phase of your Salesforce project may take anywhere between *a few weeks to many months or quarters*:

- The technical scope and the number of domains of your Salesforce solution to be deployed and released
- The geographic scope of users and regions to work with your solution

- The number of teams and users to be trained
- The skills and resources available for deploying, migrating data, and communicating with and training your users

> **Tip**
>
> In *Chapter 3, Determining How to Deliver Your Salesforce Project*, you considered how to phase your Salesforce project. If you are yet to determine this, go back to the chapter, and make your project phasing plan.

Let's see who will be overseeing the roll-out of your Salesforce solution.

The role of your Salesforce CoE in the roll-out phase

In *Chapter 3, Determining How to Deliver Your Salesforce Project*, you envisioned the structure and responsibilities of your **Salesforce Center of Excellence (CoE)**. Now it is time to execute and establish it.

While deployment and data migration seem to be technical tasks, your Salesforce CoE is ultimately accountable for overseeing those activities and managing any escalated issues that may arise.

In addition to that, in *Chapter 10, Communicating, Training, and Supporting to Drive Adoption*, you will learn how you and your Salesforce CoE will be planning and preparing change and communication plans, system cut-offs, user training, go-lives, and setting yourself up to provide hypercare and production support – all with the goal of ensuring adoption to reap the benefits of your investment.

Let's get started with the first topic!

Deploying your Salesforce solution through environments

Salesforce deployments come in many forms, and their duration can span a few minutes to many hours or even days for heavily customized, old orgs with limited technical governance.

While *what* you deploy is unique to your organization and project, the steps for *how* you deploy your entire solution from one sandbox to another – and finally to production – *at an abstract level* are as follows:

- **Plan deployment**: There are two parts to planning a deployment:
  - **Organization deployment planning**: Make sure you have aligned the following key things with your stakeholders and project team members:
    - What you want to deploy
    - When you want to deploy

- Who you expect to be part of the deployment, and what they are expected to do

- Any development freeze periods have been communicated

- **Technical deployment planning**: Determine and prepare the following items before starting your deployment:

  - **Prepare a release artifact**: Depending on your development method (refer to *Table 7.2*), this could mean preparing a Change Set, preparing the `.zip` file with metadata components, or creating/updating a version of one or multiple unlocked packages

  - **Identify and list pre- and post-deployment steps**: Throughout the sprints in the development phase, your development team should have continuously been adding any deployment steps that cannot be deployed via Change Sets or Metadata API and hence will require manual deployment steps

> **Tip**
> As mentioned, not all Salesforce metadata is deployable through Metadata API – visit the *Metadata Coverage* report at `https://developer.salesforce.com/docs/metadata-coverage` to find out whether any of your solution's components require manual deployment.

  - **Decide on a fallback plan**: Determine what you will do if the deployment fails, takes longer than planned (fails), or does not pass smoke-test validation in the target environment

- **Deploy your solution**: There are generally three parts to the technical deployment of your solution:

  A. **Execute pre-deployment steps**: These may include the following steps:

     i. Enabling various Salesforce features in the **Setup** menu. Some examples include setting **Web-to-Lead Enabled**, **Enable Web-to-Case**, **Allow users to relate a contact to multiple accounts**, and **Enable Field Service**.

     ii. Installing AppExchange apps.

  B. **Deploy the release artifact**: After having prepared your release artifact, deploying can be as easy as the click of a button (or typing an `sfdx force:source:deploy` command and pressing *Enter*). However, deploying a complete solution the first time can be hard, as it may be the first time all the components have been deployed together to a fresh Salesforce sandbox or production instance.

  C. **Execute post-deployment steps**: These may include the following steps:

     i. Execute manual post-deployment of components that are not deployable.

     ii. Creating or importing reference data, such as products, price books, price book entries, **configure price quote (CPQ)** configuration data, and Field Service Lightning configuration data.

    iii. If your solution integrates with external systems, you will likely have technical dependencies with them. This could be certificates or generating HTML snippets for a web form that your web development team needs for your website.

- **Deployment validation**: Smoke test validation of select features and functionality. You should limit creating or modifying data in production:

  - Your team will gain experience and develop a routine for how to deploy your solution as you deploy through your sandboxes to production. Develop a routine and maintain good discipline within your development team to make sure your deployment artifact and process is documented – and *critically* – updated after every sprint/release.

Tip

A common artifact (sometimes agreed to be a project deliverable by your implementation partner) is a detailed deployment plan or deployment checklist (sometimes referred to as a runbook). Your detailed deployment plan is for the deployment team to follow for each deployment and should include the following items:

- All pre- and post-deployment steps, including status, dependencies, exact timings, and ownership for each task in the plan

- Deployment timeline and stakeholder communication plan

- Description of how deployment sign-off is done by your product owner and other stakeholders

---

Next, we'll move beyond the technicalities of deployment.

Performing system integration testing

You carry out system integration testing to ensure your developed solution works as intended when deployed *all together*. What do we mean by *all together*? It can mean two things depending on the type of Salesforce project:

- **Greenfield Salesforce implementation**: If this is your first Salesforce project and you are yet to have users live in production and if your development pipeline consists of *multiple streams/ product teams or developers* with their own sandboxes and/or feature branches, merging them may cause conflict – and that's the goal of SIT – *testing the integration of your system*:

  - Catching bugs now rather than later – in UAT or production – will make for smoother UAT and, ultimately, better adoption of your solution in production

- **Brownfield Salesforce implementation**: If you already have users live on your Salesforce production instance, you want to make sure any new features or changes don't break existing processes.

You can opt for performing SIT at the end of each sprint or after the build phase.

The exact activities and steps of performing SIT will depend on the nature of your Salesforce project, however, performing SIT typically consists of the following activities:

1. **Deployment error identification and resolution**: As mentioned in the previous section, you are likely to encounter deployment errors the first time you deploy your entire solution. Resolving those deployment errors is your first step.

> Tip
>
> Do deployment dry-runs before the first production deployment of your initial (or any major) release in a staging sandbox (preferably a full copy sandbox). This can be done as part of preparing a sandbox for SIT or UAT, and it gives your deployment team an opportunity to fine-tune the detailed deployment plan to make sure no steps are missed or done in the wrong order. You should also – throughout your development phase – be executing test deployments (or deployment validations) to understand the number and nature of deployment errors. These activities will help increase your confidence that the production deployment will go smoothly by minimizing the risk of your deployment failing or taking too long.
>
> If you are overwhelmed or stuck resolving deployment errors, review this *SalesforceBen* article by Andrew Davis at `https://www.salesforceben.com/checklist-for-debugging-salesforce-deployment-errors/`.

2. **Metadata merge conflict identification and resolution**: This can consist of one of the following two things depending on your Salesforce project. Once these are resolved, your solution can technically be deployed to the next environment:

    *   **Greenfield**: Identify conflicts created when merging metadata components from different streams and resolve them

    *   **Brownfield**: The same principle as greenfield applies here, except you are identifying conflicts between the changes in your Salesforce project and what is already in production

3. **Regression testing**: Beyond technical deployment, you need to test that your incremental changes don't break any existing processes. It can be executed through manual testing or **automated testing** using tools such as **Provar** or **Selenium**. They do, however, require setup (and cost) and training, so you need to weigh the benefits versus the cost and time. Another aspect of regression testing is testing that an integrated solution's interfaces to other systems still function as before or – if any changes were modified or developed and tested in lower environments – as according to any new modifications.

> **Tip**
>
> If you have a large, long-running project, you may consider implementing automation testing as part of your development life cycle. For smaller, short-lived projects – in Salesforce setups without an existing test automation tool available – you are likely better off dedicating time and resources to manual regression testing.

Next, let's use UAT as a catalyst for adoption and a sense of ownership among your end users.

Driving UAT of your solution

You have deployed your solution to the sandbox in which you intend to carry out UAT. Great!

UAT is one of the key milestones of a Salesforce project. If your **acceptance testing team** (or customers and partners) don't accept the solution you are building, nothing else really matters.

Let's see next what you need to prepare to plan for a successful UAT.

Planning for a successful UAT

To set yourself up for a successful UAT, you need a plan. Let's look to see what it should contain:

- **Lean meetings**: These should be short meetings to share updates, and in some of them make decisions, though these are not meant for brainstorming or discussions:

  - **UAT kickoff**: Here, your project team introduces UAT to the acceptance testing team, including the duration, activities, processes, and outcomes of UAT

  - **Daily check-in/check-out**: These are quick status calls with your **project manager** (**PM**), **product owner** (**PO**), **business analyst** (**BA**), and **quality assurance** (**QA**) specialist to assess the number of reported bugs, their classification, and their resolution status

  - **UAT sign-off**: Determine how UAT is officially signed off and finished

- **What should you test in UAT**: The following are the two things you should test in UAT:

  - **End-to-end test cases**: These are a collection of test scenarios based on the user stories in the scope of your solution. The QA specialist assigned to your project, together with your BA, should create the end-to-end test cases.

  - **Non-vanilla flows**: Invite your acceptance testing team to report bugs encountered when not following the test cases' flow in the right order.

> **Important note**
> A *well-architected and well-tested solution* will consider and manage edge cases and exceptions and will guide users to the right path if they try to do something "the wrong way".

- **Who should participate in UAT**: The following resources and stakeholders are key to a successful UAT:

  - **The sponsor**: They should kick off UAT by sharing the vision for your Salesforce project and explaining how important the acceptance testing team's job and effort is.

  - **PO**: They are responsible for responding to reported bugs from the acceptance testing team. See the further point on *Define a process to follow*.

  - **Acceptance testing team**: You have already interacted with some of them (your target end-users) in the pre-development phase and throughout the development phase of your Salesforce project. Now it's time to get them together as your acceptance testing team to test your solution from end to end.

> **Important note**
> The end users of your solution whom you invite to participate in UAT may not have been part of UAT before – or may not have been part of an agile or hybrid agile project before. Managing the expectation of your UAT participants is key to a smooth process. To do so, be open and transparent about the approach you have chosen for your project and why. If you are building an MVP to bring value quickly for a limited scope, explain that. And let your UAT team know that more features are planned but are not included in the initial release being tested now in UAT.

  - It's a balancing act to decide who and how many individuals should be part of this team. By involving too few end-user representatives, you risk not testing comprehensively enough, which could lead to a suboptimal, bug-prone solution that is poorly adopted. On the other hand, having too many people on the testing team can lead to you spending a lot of time and effort on upfront training and instructions – let alone handling the greater volume of reported bugs:

    - As a bare minimum, you need to have at least one test member for each persona – in Salesforce, this could (from a system-perspective) be represented by a combination of **record access grants** (`Role`, `Territory Assignments`, and `Queue Memberships`) and **system permissions** (`Profile` and `Permission Set Group`).

> **Tip**
>
> To lay the groundwork for user training and to begin recruiting change agents, you should consider having at least one representative per domain in the scope of your solution (Sales, Service, Marketing) *for each* planned geography (region/country). If your solution is tailored to different subdomains or subdepartments, for example, telesales and key account management, which have different processes to test, you should have at least one representative per subdomain.

- **BA and/or QA specialist**: Triage, assess, and categorize reported bugs and also potentially propose solutions to fix actual bugs

- **Development team**: Fix identified and prioritized bugs

- **Define a process to follow**: Establish and communicate how to report a bug and the process to triage and potentially fix the bugs:

  - Share the bug template with all acceptance testing team members. The bug template should – when correctly filled out – make it easy for the BA or QA specialist to triage, reproduce, and classify a bug. It should contain the following information as a minimum:

    - The username of the bug reporter

    - Test case name or ID

    - At what step of the test case the user encountered the bug

    - What the user expected to happen versus what actually happened

    - A link to the record/page

    - Screenshots of the entire browser window, including the URL bar, which sometimes provide the project team with valuable context about the issue and reduce back-and-forth communication with the tester

- **Determine bug classifications and actions for each**: These typically include the following classifications but may vary depending on the nature of your Salesforce project:

  - **Change request**: Your PO should assess and address these by either communicating to the bug reporter that the request is considered and put in the backlog, or that it is not considered and why.

  - **Not a real bug** This could either be due to user error or faulty data. In either case, there is no development required to fix the issue. Your BA or QA specialist should show the bug reporter how they execute the test case and what data to set up beforehand.

  - **Bug**: The BA and developers can assess the bug, determine the reason for the behavior (root cause), find solutions to fix it, and estimate the effort to do so. The PO can review and determine the prioritization of the bug against the other bugs in the backlog.

> **Tip**
>
> Many bugs reported during UAT are caused by wrong user setup (profiles permission sets, queues, and so on), because test cases are not followed, or because of **faulty test data**.
>
> To minimize the risk of bugs being reported due to these causes, make sure you provide adequate instructions to your testing team and double-check that test data is correctly set up and ready for the testing team. Test data includes the following:
>
> - Salesforce test user accounts to log in with
> - Master data, such as accounts and contacts
> - Transactional data, such as opportunities, orders, and cases
> - Reference data, such as products, price books, and price book entries

Through the course of UAT, you will likely receive a lot of feedback or bugs from your acceptance testing team. Be open to feedback and thank the team for their effort.

> **Tip**
>
> Srini Munagavalasa has written an entire chapter dedicated to UAT in a Salesforce context. I suggest you check out *The Salesforce Business Analyst Handbook* (`https://www.amazon.com/Salesforce-Business-Analyst-Handbook-techniques/dp/1801813426`).

Next, let's make sure you are ready to get your data into Salesforce!

Checking your data migration preparedness and signing off to go live

You are now in the final stages of the development phase. You have successfully deployed your Salesforce solution and executed SIT and UAT.

In this section, we are going to highlight the key practical considerations to evaluate whether you are really ready to start migrating data into your Salesforce production environment.

Evaluating your data migration preparedness

Before you embark on migrating your data to production, make sure you have prepared – and practiced – your data migration plan. It should have been done in the development phase.

> **Important note**
>
> If you haven't prepared your data migration plan, you need to pause and check out the *Preparing your data migration plan* section in *Chapter 7, Building and Testing Your Initial Release*, and get cracking before proceeding.

Great, you have your data migration plan in place! This means you have determined how you will technically get your data out of any legacy systems, other applications, and sources, and you know how and who will be transforming and loading it into Salesforce.

Let's look at the list of things you can use to evaluate your preparedness to begin migrating your data into Salesforce:

- You have executed successful data migration dry runs in a staging sandbox.

- You have confirmed that your Salesforce solution is deployed to production – including its configuration data.

- You have confirmed with the development team – beforehand – whether automation and validations are designed to handle migrated data or whether they need to be made inactive during data migration. If the latter, you have confirmed who will do so and when.

- You have aligned with and confirmed the availability of IT and business stakeholders to verify and validate your data migration in production. The same goes for the source system experts to be available in case of issues that need resolution by a new source system extraction run.

- If your implementation partner will participate in migrating your data, make sure you comply with any laws applicable to the countries you operate in according to the nature of the data they will have access to and process.

- You have agreed and aligned the amount of time you have for migrating to production – and validated this to be possible in at least one dry run.

- You have a rollback plan in place. No one wants a data migration to fail, but the truth is some data migrations do fail or take longer than expected and acceptable by your stakeholders. So save yourself a headache and determine what you will do in the event the migration cannot be signed off.

- You have communicated the data migration timing and impact on any system downtime to the teams who are about to go live.

> **Tip**
>
> Read *Salesforce Data Architecture and Management Data* by Ahsan Zafar (CTA) to understand and fill in the details of your data migration strategy, plan, and execution.

You have now migrated your data to Salesforce and wish to proceed with training your users and start having them use Salesforce to deliver better experiences for your customers, partners, and employees.

Before you let your users into Salesforce, let's do a final quality check.

Signing off to go live

You will be ready to let users go live in Salesforce when you can tick off the following items:

- User acceptance testing was completed with a successful outcome.

- Your Salesforce solution is deployed to production and is verified.

- Your data migrated to Salesforce is validated by both your IT team and the business stakeholders who are going live.

- **User training** is completed for the people about to go live. To minimize disruption to business operations, training sometimes happens in parallel with deployment and data migration. We'll cover user training in *Chapter 10, Communicating, Training, and Supporting to Drive Adoption*.

- Your users are **provisioned** for your Salesforce production org.

Well done! You have reached the end of deploying your release and migrating your data to Salesforce.

Let's wrap up this chapter!

Summary

In this chapter, you have covered a lot of material. We began by guiding you through the key activities and milestones in the roll-out phase. Next, we covered the planning and steps for deploying your Salesforce solution through environments.

You learned how to perform SIT and UAT to make sure your solution works as intended from end to end and is accepted by representatives of your target end-users. Finally, we covered how to evaluate your preparedness for migrating your data to production and provided a checklist for going live.

Having deployed your Salesforce solution to production, you are now ready to train your end-users to set them up for success and to communicate with and support them as they learn to use your solution. All that and more to come in the next chapter of your Salesforce implementation.

10

Communicating, Training, and Supporting to Drive Adoption

In this chapter, we will focus on preparing your **change management plan template**, as well as planning your **roll-out** activities in **phases** or **waves**. As part of this, you will reiterate what will change as your Salesforce solutions replace existing ways of working – and prepare your organizational enablement plan to manage the changes. We will introduce various training models as well as and what to consider when using each of them. Then, we'll cover how to provide hypercare in the initial period after go-live and how and when to transition to production support operations. Next, we will discuss how to monitor and drive adoption. Finally, we will look at how you can begin to evaluate and compare the benefits achieved with your original business case.

You will learn to design a change management and training plan for your roll-out. Then, you will learn how to provide efficient hypercare support for users going live with your solution, and how to monitor and drive adoption.

We will follow our scenario company, **Packt Manufacturing Equipment** (**PME**), as it progresses through the roll-out of its Salesforce solution.

By the end of the chapter, you will understand the critical activities you need to plan and execute to ensure the business value is unlocked by getting your users working in Salesforce.

This chapter will cover the following main topics:

- Planning, communicating, and managing change
- Providing hypercare and managing production support
- Monitoring and driving the adoption of your Salesforce solution

Let's go!

Planning, communicating, and managing change

The purpose of the following activities, in short, is to ensure that the business value you wish to achieve with your investment in Salesforce is unlocked, realized, and maximized.

Let's get started!

Understanding the scale of change to come

In *Chapter 3, Determining How to Deliver Your Salesforce Project*, in the pre-development phase of your project, you identified the actors who would be affected by the change as a result of your Salesforce solution. A lot has happened since then. You have progressed through the pre-development and development phases, and many of the assumptions and hypothetical processes and solutions may have changed – to better fit your business needs.

What you should ideally have done throughout the development phase is work with a **business change consultant** to identify how the new solution and processes differ from current ways of working, and for whom.

Here are some of the typical stakeholders who are affected by the change as a Salesforce solution is implemented:

- The end users
- Managers of end users
- The leadership of local organizations
- Global leadership
- Application management staff
- Data, BI, reporting, and analytics staff

After you have identified the stakeholders who will be affected by the change, let's move on and prepare for the change to come!

Preparing your change management plan template

Traditionally, go-live is referred to as the point at which your users log in to Salesforce production for the first time and perform work in it.

But what is the value of that? There isn't any necessarily. Going live itself does not equal adoption, and certainly is not the assurance of the best, correct use of your solution.

The traditional definition of go-live may be semantically accurate, however, from a change management and business value perspective, most of your future end users have never seen or worked with your solution or know about your solution's capabilities, limitations, and potential.

The change *will* happen for your stakeholders as you roll out your solution. Make sure you have a plan for managing it – before the change occurs. Keep your plan simple and update it as you learn from the roll-out.

In *Figure 10.1*, you can see a consolidation of the many activities you need to consider and plan for go-live to make the onboarding and the change journey as smooth as possible – to ultimately reap the benefits of your Salesforce investment. The numbers at the top represent the weeks relative to go-live:

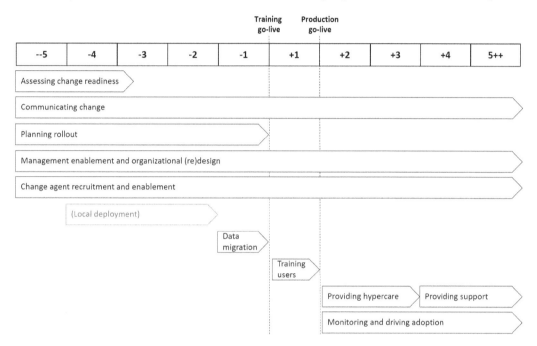

Figure 10.1 – Change management activities relative to go-live

Let's go through each activity:

- **Assessing change readiness**: Perform interviews with a limited number of people from the different stakeholder groups to assess their sentiment, sense of preparedness, and level of excitement or concern about changing to a new way of working in a new system. Depending on the assessment, initiate appropriate actions and communicate accordingly.

> **Tip**
>
> Review the *Salesforce Change Readiness Assessment* document from **The Salesforce Way** at https://www.salesforce.com/content/dam/web/en_us/www/documents/ services/the-salesforce-way-change-readiness-assessment.pdf.

- **Communicating change**: Create a plan to communicate in the context of the user group(s) going live. You should engage with local leadership to understand the context and circumstances that affect the organization – also refer to the results of your change readiness assessment:

  - Your comms plans should include *what* to communicate *to whom*, *when*, and from *which sender* (ideally, the local leadership or **change agent**, also called the **champion**). Here are some examples of what you can communicate:

    - *Before training*: Build excitement yet manage expectations by not overpromising.

    - *During training*: Key aha moments of joy from the training sessions.

    - *After training*:

      - Success stories from users. Tips such as "did you know you can do or see this in Salesforce?"

      - Salesforce productivity tips if you have configured or enabled any of them – **Email Templates**, **Quick Text**, or **Macros**.

      - Responses to bugs reported/new features requested.

      - Key adoption metrics, such as top and bottom adoption by end user group.

  - In your communication, maintain a healthy balance of creating excitement and managing expectations – people are complex beings, able to hold multiple feelings at the same time: fear, excitement, eagerness for change, and, at the same time, resistance to it.

    - Remember the purpose and vision of your Salesforce project that you established in *Chapter 1, Creating a Vision for Your Salesforce Project*.

    - In your communication, continuously answer this key question for your user groups: "*What's in it for me?*"

Important note

Be sure to communicate with *all* stakeholder groups – and tailor the communication to each group and use the vehicle (i.e., the comms channel) that's most appropriate.

Bring representatives of your stakeholders together with a business change consultant and an adoption and communications specialist to create **persona-specific** communications for each stakeholder group. This will enable you to better understand and empathize with your stakeholders and to craft targeted, relevant, and effective communications for maximum impact.

- **Planning roll-out**: This entails identifying the individual users by the persona your Salesforce solution involves, in addition to all the resources you need to be involved in carrying out the activities shown in *Figure 10.1*. You need to plan the timing of the activities to ensure everyone is available. Be mindful of any language considerations.

- **Management enablement**: Enablement and buy-in from management is critical to the success of your roll-out. You should enable managers to deal with the resistance and concerns their team members may encounter as part of the roll-out and use of your Salesforce solution. This consists of educating the management team about the purpose and capabilities of the Salesforce solution, as well as handling any general fear of change that may arise – and how managers can help alleviate it. After initial training, managers of users are key to helping support and encourage the use of your solution.

- **Organizational (re)design**: At the beginning of this section, you identified the stakeholders who will be affected by the change. The extent of the change depends on the nature of your Salesforce solution:

  - Your solution will require your staff to work with a new system (Salesforce). This change can be managed with adequate training (we'll cover this in the next section).

  - However, if your new solution requires users to work in a new way (in terms of the processes) that is radically different and more focused on value-adding activities rather than rudimentary data entry and updating, this change may require new skills and competencies that your staff may or may not have – and may be more or less susceptible to acquiring:

    - Identify the gaps in the current staff's capabilities and create the necessary training plan (non-Salesforce).

    - Update role descriptions to make sure future hiring for those positions accounts for the new skill sets required.

    - For any current staff who don't possess the skills required in the future and are not susceptible to acquiring them, create a restructuring plan to find alternative opportunities for them within your organization.

- **Change agent recruitment and enablement**: As mentioned several times, you should have involved end user representatives throughout your Salesforce project – to understand their situation and design and build a solution that improves their daily work. Leveraging and empowering **change agents** is an excellent method to help end users with questions – because they know what considerations have gone into making the solution work the way it does.

> **Important note**
>
> Recruit and enable your future change agents early in your project. As you can see in *Figure 10.2*, the journey to becoming an effective change agent begins in the pre-development phase and continues through development and roll-out. 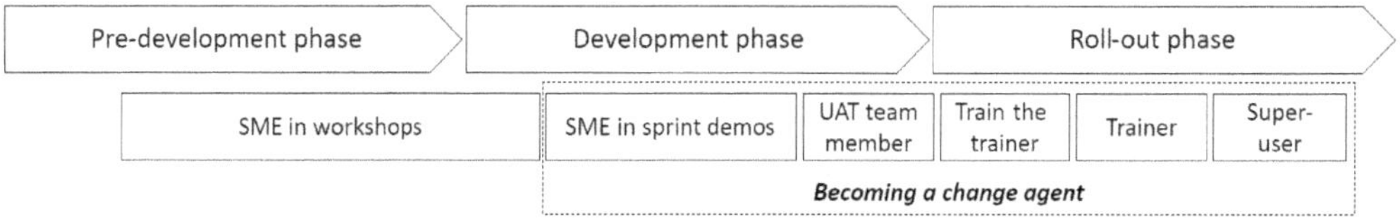
>
> | Pre-development phase | Development phase | Roll-out phase |
> |---|---|---|
> | SME in workshops | SME in sprint demos | UAT team member | Train the trainer | Trainer | Super-user |
>
> *Becoming a change agent*
>
> Figure 10.2 – A change agent's journey

- **Local deployment**: If you are rolling out to multiple geographies, you may have decided to create an "MVP template," which will require modification to be ready for local use. This typically includes the following:

  - **User and localization setup**: Setting up a configuration for things such as the following:

    - **User management and access**: The local **Role Hierarchy**, **Queues**, **Public Groups**, and **Sharing Rules** configuration, local currency setup, and of course, creating and provisioning users with the current **Profile** and **Permission Sets** and user locale settings.

    - **Localization**: You may have created a solution that allows for *some* process variation to be supported. That may require local configuration such as **Custom Settings** and/or **Custom Metadata Type** records to hold reference configuration data and also translations.

  - **Local integrations**: If your local organizations have separate backend systems or other systems that need to integrate with Salesforce, you need to start up the analysis of this already in the development phase of your Salesforce project.

- **Data migration**: If you are rolling out to different local organizations you need to create and align the data migration plan for each of those roll-outs.

- We'll cover the remaining activities shown in *Figure 10.1* – **Training users**, **Providing hypercare**, **Providing support**, and **Monitoring and driving adoption** – in the following sections.

Many of the activities in this chapter of your Salesforce project are happening in parallel, so keep all the plates spinning to progress toward your goal. A wise woman once said, "perfect planning prevents poor performance", and that certainly applies to roll-outs.

Let's dive into how you can roll out in waves!

Phasing the roll-out of your Salesforce solution

While you can reuse many elements of your change management template for further roll-outs, you need to execute all the activities *for each* wave of the roll-out. Therefore, it makes sense to group or cluster some local organizations together in a **roll-out wave**.

A phased roll-out approach has many benefits, but there are some important considerations for choosing it. The main consequence of choosing a phased roll-out approach is **concurrent systems**. Let's first consider the key reasons why you might need to have concurrent systems in the roll-out phase:

- Your initial Salesforce release (the MVP) does not support all capabilities currently covered by your legacy system

- Different geographies will roll out your Salesforce solution at different points in time, resulting in different geographies working with different systems

- Different functional user groups (marketing, sales, and customer service) will roll out your Salesforce solution at different points in time, resulting in different functional user groups working with different systems

There may be many reasons why and instances in which you will need to keep two systems active. Whatever the reason, concurrency may have the following impacts, depending on the nature of your Salesforce solution and roll-out planning:

- Users on both systems at the same time means they will be update and entering new data into both systems and it will be out of sync:

  - This can be mitigated by doing regular delta data migrations or (if a viable economic option) creating interim integrations between the two systems until the old system can be retired:

    - The old system sometimes is left available but in a read-only state so users can reference legacy data if needed. If that approach is taken, it should be limited to a short period after go-live.

- Another consideration is the feasibility of integrations with third-party systems:

  - Can the new Salesforce system and legacy CRM both be integrated with other systems at the same time?

    - If not, choosing whether Salesforce for your legacy CRM should be connected to the third-party system is an important decision to be made.

In *Figure 10.3*, you can see an example of how our scenario company, **PME**, has chosen to phase the roll-out of their Salesforce solution:

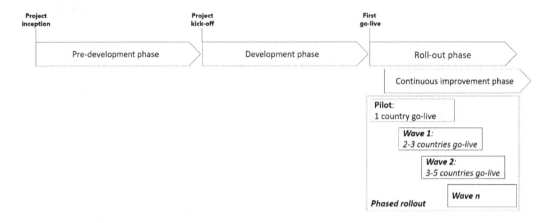

Figure 10.3 – Example of a phased roll-out overlapping with continuous improvement

PME has decided to go live initially only with one country – the pilot country.

Their approach has three main advantages. First, it requires less effort (the first country has fewer people who are going to use the solution) for the project team to plan and execute all the roll-out activities. Second, it allows the project team to collect valuable feedback on the solution, which can be used to prioritize changes as the Salesforce program enters the continuous improvement phase. Third, the project team can test the change management and roll-out plan – and can optimize accordingly ahead of further roll-outs.

As the PME Salesforce project team rolls out to the pilot country, they optimize their approach to be able to industrialize it – enabling them to accelerate roll-outs in increasing waves, with more countries per wave.

> **Tip**
>
> Align with your company's leadership team or **project management organization** (**PMO**) to understand what constraints or concurrent projects or initiatives may impact the local organizations to which you wish to roll out. Include these considerations when preparing your phased roll-out plan.
>
> Additionally, you should critically work with the leadership of the local organizations to align and confirm the change management and roll-out plan for their organization.

Let's take a closer look at user training!

Training your users

User training is a key milestone in your Salesforce project – and if you have prepared your stakeholders well for the change to come, it can be a cause of celebration to finally be able to use Salesforce!

Training is the change management activity that required the most effort and has the greatest potential for impact in a short period, so be sure to plan and prioritize it accordingly.

In the following table, you can compare different training methods to consider what would work best for your Salesforce project:

Training method	Training type	Preparation effort	Effect	Appropriate use
Direct in-person training	Instructor-led	High	High	Initial end user training
Direct virtual training	Instructor-led	Medium	Medium	Cheaper alternative to in-person training. Good for training supporting roles (BI/reporting staff, HR, and finance).
Train-the-trainer	Instructor-led	Medium	N/A	Ahead of and in combination with direct end user training
Peer-to-peer training	Peer-led	Medium (not yet scalable)	High	A great method for direct end user training follow-up and for training new staff and staff who are changing roles
Training videos	E-learning	Medium	Medium	Supplement to direct end user training
Salesforce Trailhead	E-learning	Low	Medium	For your Salesforce admins and production support staff
Salesforce MyTrailhead	E-learning	High	Medium	A great method for direct end user training follow-up and training new staff and staff who are changing roles
In-app guidance	On-the-job training	Low/Medium	Medium	Supplement to direct end user training

Table 10.1 – Common training methods and appropriate use

Leveraging the train-the-trainer model, in combination with direct enduser training, can accelerate training, as it allows you to run multiple training sessions concurrently.

> **Important notes**
> - People's affinity for learning methods differs, so include a mix of training methods in your training plan.
> - If you're rolling out an initial release of a Salesforce solution to people who have never used it before, you must provide one form of direct end user training or another.
> - Regardless of which methods of training you opt for, you need to make sure local leadership is present in the training sessions. The executive sponsor of your Salesforce project (or another member of senior management) should introduce the vision and objectives of the new Salesforce solution before diving into system training.
> - If you include videos in your mix of training methods, be sure to keep them short (~30-120 s) by having each video cover small steps in an end-to-end process.
> - You can consider creating one-pagers (sometimes referred to as **cheat sheets**) for each actor with key topic training areas.
> - Make sure you have a training materials governance process in place to continually update these materials as your solution evolves.
> - Make sure to plan and execute training for your future Salesforce admin(s) and wider IT organization:
>   - This includes application management staff, as well as Data, BI, reporting, and analytics staff.
>   - Provide your future Salesforce admin(s) and application management staff with training on Salesforce administration and the clouds in the scope of your Salesforce solution, and encourage and support them to become certified.
>   - Hand over the solution in your initial release to your application management staff. In addition to the processes supported by the solution, focus on the user management and maintenance required.

Next, we'll look at how you can support your users after training and going live!

Providing hypercare and managing production support

The period after user training is key and fragile. As such, a heightened level of support is required to be able to quickly respond to questions and issues faced by end users. That phase is called **hypercare** – let's dive into it!

Providing hypercare

Hypercare is a phase in your Salesforce project as much as it is a capability.

Let's break the capability into its parts:

- **The hypercare process**: When your users begin working in Salesforce, they may get stuck, face issues, and require further assistance beyond the training and training materials you have provided. Be sure to communicate the process your users should follow to get help. It may be similar to the process you used for SIT and UAT, so use that as a starting point.

  - Depending on your setup and available resources, you need to determine what levels are involved in your hypercare process. Ask your end users to state the priority/severity when submitting bugs so your hypercare team can prioritize accordingly:

    - Depending on your project setup and the contract with your implementation partner, it may be helpful to define a specific **service-level agreement** (**SLA**) for each defined severity so your end users know by when to expect their bugs to be reviewed and/or resolved.

  - You want users to submit issues (potential bugs) using a pre-determined template (revisit the *Driving UAT of your solution* section of *Chapter 9, Deploying Your Release and Migrating Data to Production*).

  - Similarly, you should also have a process for how users communicate new ideas or change requests:

    - Depending on the maturity of your organization and end users – that is, whether they would struggle to distinguish a bug from a change request – you may opt to only have one template for them to submit both bugs and change requests.

- **Hypercare people**: Who receives the submitted bugs can vary; it may be your implementation partner's *business analyst*(s), your own *Salesforce application management team*, or your *product owner*. Ideas should be received and reviewed by your product owner, and the outcome communicated back to the reporter. Here are the different responsibilities in hypercare and what roles take them on in order to fix qualified and prioritized bugs:

  - You will need to have functional consultants, developers, and QA specialists available (at least one of each) – how many of each resource depends on the nature and size of your Salesforce project.

  - You need to have your PO(s) fully available – one for each of the business domains of your Salesforce solution going live.

  - It is also best practice to have architects (or at least one) available in case a fix requires a more complex redesign of solutions.

- **Hypercare systems**: You should also determine what system your users should use to submit bugs and ideas:

  - For bug reporting, you can use Salesforce `Cases` (this requires configuration to fit your organization and process) or your help desk ticking system. It is important that triaged bugs that need resolution should be registered and tracked in the system you use to manage development activities – for example, *Jira* or *Azure DevOps*.

  - For new ideas, you can utilize Salesforce's `Case` object (requiring configuration to fit your organization and process), which allows users to create Case records in Salesforce that you can assess and respond to.

- **Hypercare data**: From these systems, you should monitor and analyze the volume, trend, and resolution of bugs and ideas.

> **Tip**
>
> Rather than obsessing over isolated user feedback, criticism, and resistance to change, have a balanced approach to user feedback; take note of valid concerns and improvement suggestions – and importantly – *respond patiently and empathetically to all user feedback*!
>
> Your organization and implementation partner will become more capable to support new users during hypercare as you progress with the roll-out of your Salesforce solution to more users.

Having covered what hypercare is and how it is carried out, let's discuss how and when to progress to ongoing support!

Transitioning to ongoing production support

Hypercare is more expensive than regular ongoing production support. As such, it may be tempting to quickly transition to production support once hypercare bugs begin to decrease.

Recall our definition of a successful roll-out from the previous chapter – to maximize value from your investment in Salesforce:

> *A successful roll-out phase should be defined as the point at which your organization, customers/partners, and users use and gain value from your Salesforce solution, and it delivers the target business outcomes you laid out in Chapter 4, Securing Funding and Engaging with Salesforce and Implementation Partners.*

You should consider the following four points before deciding to transition from hypercare to ongoing production support:

1. Have you *defined and aligned your process* for ongoing production support and allocated sufficient resources to it – proportional to the nature of your Salesforce solution, its required maintenance, and the number of users living in it?

2. Is the *team* that will be responsible for ongoing production support…:

   - Going to be an in-house team or staffed by an external partner as a managed service?

   - Trained in your Salesforce solution?

   - Onboarded to the bug and idea process?

> **Important note**
>
> The team you have allocated to hypercare will likely include some of the same members of the development team of your Salesforce solution. They are experts on your Salesforce solution and therefore the best and quickest actors to reproduce bugs, troubleshoot, and fix them. If you transition prematurely, likely, some of these resources will be allocated to other projects.

3. Are you reaching your *adoption* targets?

   - Critically, read the next section before considering transitioning to ongoing production support.

4. Are you beginning to *achieve the business results* you laid out in the business case for your Salesforce project?

Unless you can confidently say a resounding *yes!* to all four points, you should hesitate to transition from hypercare to ongoing production support.

Next, let's turn to the important topic of adoption!

Monitoring and the driving adoption of your Salesforce solution

In parallel with providing hypercare and support, you should closely monitor and try to understand the adoption of your Salesforce solution.

As mentioned in *point 4* in the previous section, you shouldn't consider transitioning from hypercare to ongoing production support before you reach your adoption targets. Why not? If your solution isn't adopted in the hypercare phase of your roll-out, how will you make sure it is adopted later and whose responsibility will it be? When your project enters the continuous improvement phase, the focus for many stakeholders will shift to new feature development, and for your roll-out/local deployment team to prepare for the next wave.

For these reasons, I would propose that you – as the one accountable for your Salesforce implementation – expand the scope of what hypercare entails, and when it may finish.

Let's move beyond simple quantitative bug reporting and login metrics!

Defining and measuring adoption

According to Oxford Learner's Dictionaries, the verb **adopt** has the following meaning within the context of adopting a method:

"Adopt something to start to use a particular method or to show a particular attitude towards somebody/something."

We could translate that definition to the following in a Salesforce context:

"Enabling users to perform work in Salesforce and to perceive it to be a valuable improvement for their work life."

Let's break down the definition further into its two parts, and some of the key ways you can measure and track adoption in a Salesforce context:

1. **Perform work in Salesforce**: You need to track whether your intended users are using your solution – and to what effect. This involves tracking two parameters:

   - Logins:

     - Daily, weekly, and monthly

   - Usage:

     - **Efficiency**: The number of records created – for example, **Leads**, **Opportunities**, **Cases**, and **Tasks** measured per timeframe (hour, day, or week).

     - **Effectiveness**: This could include the number of converted **Leads**, **Opportunities** closed (won/lost), and **Cases** closed per timeframe (hour, day, or week).

> Tip
>
> The specific metrics relevant for you depend on the nature of your Salesforce solution and what users (or customers and partners) it is intended for. Work with your change and adoption consultant to define which metrics are appropriate for you. You can measure and track (most of) the quantitative, objective adoption metrics using the free Salesforce Adoption Dashboards by Salesforce Labs:
>
> ```
> https://appexchange.salesforce.com/
> appxListingDetail?listingId=a0N30000004gHhLEAU
> ```

2. **Perceive Salesforce to be a valuable improvement**: A synonym of adopting is *embracing*. You should try to understand your users' sentiments and how deeply they embrace your Salesforce solution. Some good methods for doing this include surveys and interviews:

   A. **Survey**:

      i. Ask your users the classic **Net Promotor Score** (**NPS**) question, *"How likely are you to recommend using Salesforce to one of your colleagues?"* or a variation, *"How likely are you to recommend using Salesforce to someone in your network?"*

      ii. Ask for comments – for example, *"What's the best thing and the worst thing about the solution?"*

      iii. Ask whether they have noticed any improvements in the solution lately.

      iv. Use the end of the survey as an opportunity to remind your users how they can get help, report bugs, or make suggestions.

Tip

The trend of the responses to your surveys will provide great insights. Send out surveys weekly the first few months after go-live, then once every 1-3 months.

   B. **Interview**:

      i. As many facets and factors influence a quantitative measure of adoption and the answers to surveys, you should try to understand what the underlying reasons may be.

Tip

Not all people are comfortable submitting a bug if they hit a snag in your system. It still harms their adoption; you are just not aware of it. To avoid this blindside, make sure you regularly (daily during hypercare) check any **Flow Error** emails (these are sent to the email of the admin who last modified the flow or the Apex exception email recipients) as well as **Apex Exception** emails (you set the email addresses to receive these emails in Setup).

If you are receiving complaints about page loading times or general performance issues, set up reports to track key metrics, such as **Experienced Page Time** (**EPT**). Check out the *Measure Performance for Your Salesforce Org* help article, which describes how you may set it up, at https://help.salesforce.com/s/articleView?id=sf.technical_requirements_measuring_ept.htm&type=5.

You should share the adoption metrics with your executive sponsor, local country leadership, and change agents. Engage with your change and adoption consultant to identify actions and initiatives to help improve the adoption of your Salesforce solution.

Setting targets for adoption and evaluating your original business case

It should be your ambition to have a CRM system that is not only used, but also embraced and perhaps even something your employees are proud of. Providing superior solutions and user experiences not only raises employee satisfaction and retention rates but it's also great for business!

Now, Rome wasn't built in a day, but you should begin to see a trend toward a positive sentiment regarding your Salesforce solution.

For the quantitative, objective usage metrics, you need to relate your targets to the KPI targets you defined as part of your original business case in *Chapter 4, Securing Funding and Engaging with Salesforce and Implementation Partners.*

As hypercare progresses, you should – after an initial drop in performance due to your organization beginning to use a new system – begin to see your business performance improve. How quickly after go-live (for example 1, 3, 6, or 12 months) you begin to achieve the business results you laid out in the business case for your Salesforce project depends on the nature of your solution, what level of sophistication you are migrating from, and, of course, to what extent you have carried out your Salesforce project *by the book.*

Summary

Having read this chapter – and created your own change management and roll-out plan – you have set your Salesforce project up for a successful initial go-live, as well as successful further roll-outs. You prepared and executed your organizational enablement plan to manage the scale of change to come. You determined your overall training plan and determined which methods of training you will offer your users. You have learned how to provide hypercare after go-live and when to transition to ongoing production support. Finally, you have learned how to monitor and drive adoption, and you began to evaluate how the benefits compare to those of your original business case for your Salesforce project.

In the next chapter, we'll discuss typical issues you may encounter when rolling out your Salesforce solution, and how to mitigate and overcome them. You will also be provided with a checklist to evaluate the state of your Salesforce project in the roll-out phase.

11

Common Issues to Avoid in the Roll-Out Phase

The previous two chapters covered the activities in the roll-out phase of your Salesforce project. In this chapter, we will discuss some of the common issues often faced in that phase of a Salesforce project. We will also cover strategies to mitigate the issues or prevent them from arising altogether.

In *Chapter 5, Common Issues to Avoid in the Pre-Development Phase*, you were introduced to **root cause analysis**, and we'll continue using that method to dissect the common issues faced.

At the end of this chapter, you are presented with a set of **checklists** to evaluate the state of your Salesforce project as you wrap up the roll-out phase of the project.

This chapter will cover the following main topics:

- Common issues in the roll-out phase and their root causes
- Evaluating the state of your Salesforce project

Let's go!

Common issues in the roll-out phase and their root causes

Let's get started by looking at some of the most common issues for being unable to progress or complete the roll-out phase – illustrated with a fishbone diagram:

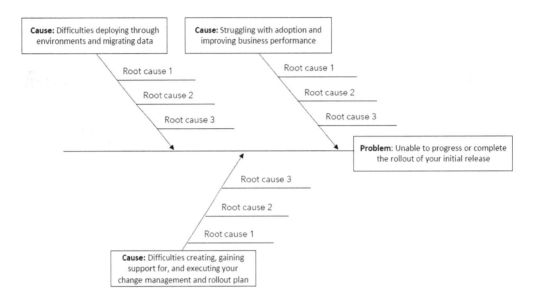

Figure 11.1 – Fishbone diagram of common issues in the roll-out phase

Let's discuss the possible root causes and strategies for the mitigation and prevention of each of the common issues.

Difficulties deploying through environments and migrating data

In *Table 11.1*, you'll find possible root causes for why you may find your project experiencing difficulties deploying through environments and migrating data – along with strategies for mitigation if you are already facing that issue:

Possible root cause	Strategies for mitigation and prevention
Insufficient *resources* available for deployment and SIT.	Ensure well in advance that the required internal and partner resources are allocated and booked for sessions and activities they are expected to participate in to carry out deployment and SIT.
Inadequate *planning* and *execution* of deployment, UAT, and data migration.	Planning and alignment on roles and responsibilities is a prerequisite for successful deployment, SIT, UAT, and data migration. It requires your PM, BA, QA specialist, and **technical architect** (**TA**) to (continue to) work closely together. While your BA, QA, and TA should know all the details and contingencies required to execute, it is your PM's responsibility to make sure all involved resources are aligned, in agreement, and aware of their respective responsibilities. Prior to attempting to deploy your Salesforce solution to production and migrating data, make sure you have a detailed deployment plan/checklist (aka Runbook) as well as a data migration plan. Refer to *Chapter 9, Deploying Your Release and Migrating Data to Production*, for more details.

Changes to or deployment of systems with which your Salesforce solution is integrated causes conflict with your Salesforce deployment.	You can prevent this issue by having awareness of other initiatives within your organization that may impact your project and cause dependencies with your Salesforce solution. Ideally, your agile team is a **cross-functional team** with knowledge and expertise in technologies and systems related to your Salesforce solution. We'll discuss cross-functional teams further in the next chapter.
	If you find yourself in the middle of a deployment to either SIT, UAT, or – worst case – production, you need to quickly understand the nature of what is different in the target environment compared to the lower environment. This is best done by your Salesforce technical architect and Salesforce integration developer, together with the same roles for your external system, along with your QA specialist. Typically, when identifying the technical root cause for the conflict, the solution to solve the issue will become apparent.
	The tricky part is often identifying the root cause and, depending on which deployment (SIT, UAT, or production), the issue has different levels of criticality. Ideally, you identify the issue in SIT (either as part of your development sprints or in a dedicated SIT after development), in which case you can proceed as described in the beginning. Similarly, if you are deploying to your UAT sandbox (or executing UAT). Here, however, there is more pressure as users are typically allocated and planned to carry out UAT, and later, training. This means you need to quickly understand how long a fix will take, and assess your need to change your plan. If you're deploying to production, you need to assess whether the time to fix impacts your overall deployment and data migration plan – and ultimately, your go-live date.

	This would compare to running a marathon without any training or – in a closer context – putting configuration and code into production without testing.
	If you are already experiencing data migration issues due to this root cause, you need to decide on one of the following courses of action:
You have failed to create a solid data migration plan and haven't performed any (successful) data migration dry runs.	• Push through with the ongoing data migration – with the enormous risk of it being insufficient – and you would have to spend significant time and resources (money) to perform a large number of data patches after going live.
	• Postpone your go-live and data migration, allowing you time to perform dry runs to perfect your data migration plan.
	From experience, the second option will (almost) always be the better option in the long term.

Table 11.1 – Possible root causes for difficulties deploying through environments and migrating data

Let's look to the next common issue in the roll-out phase.

Difficulties creating, gaining support for, and executing your change management and roll-out plan

In *Table 11.2*, you'll find possible root causes for why you may have difficulties with change management and roll-out of your Salesforce project – along with strategies for mitigation if you are already facing that issue:

Possible root cause	Strategies for mitigation and prevention
Difficulty creating your change management template and communications plan.	Don't let perfect get in the way of good – start small, experiment, and adapt and refine your change management template and communications plan as you learn.

You included an insufficient budget for change management in the business case for your Salesforce project.	Unfortunately, this is more common than not, as many organizations view Salesforce implementations as buying licenses to a standard SaaS system and provisioning users. That's a dangerous myth. The fact is, as you roll out Salesforce across your organization, change will happen. The effort and resources you put into managing that change means the difference between degrees of success (maximize value from your Salesforce investment). Worst case, you risk simply failing to get your organization behind the change and adopt Salesforce. If your organization is serious about implementing and rolling out Salesforce, treat the project as equal parts of IT and organizational change transformation.
Difficulties planning your roll-out waves.	Try to group or segment your countries by the number of users, revenue, and digital maturity. Start small with one pilot country with a medium number of users and digital maturity. Then, phase your roll-out in increasing waves as you are able to industrialize your local deployment and data migration execution, and refine your change management template and communications plan. Refer to the *Phasing the roll-out of your Salesforce solution* section in *Chapter 10, Communicating, Training, and Supporting to Drive Adoption.*

Table 11.2 – Possible root causes for difficulties creating, gaining support for, and executing your change management and roll-out plan

Let's jump to the next common issue in the roll-out phase.

Struggling with adoption and improving business performance

In *Table 11.3*, you'll find possible root causes for why the adoption of your Salesforce solution may be low and why you may not be hitting the target of your original business case – along with strategies for mitigation if you are already facing those issues:

Possible root cause	Strategies for mitigation and prevention
Your change management efforts have not sufficiently managed to do the following: 1. Inspire and *motivate* your users to embrace the positive change your Salesforce solution is bringing to them – manifested in the vision (the why) for your Salesforce project 2. *Alleviate* your users' *fear* of the change your Salesforce solution is bringing.	Learn from your change management efforts and the feedback and insights you gain as you roll out. Address each point in the following ways: 1. Adapt your change management approach and communications to focus more on the improvement your Salesforce solution will bring to your users' work life – for example, it may reduce manual work allowing them to focus on more stimulating and fulfilling (and value-adding) work. Refer to the vision you established at the end of *Chapter 1, Creating a Vision for Your Salesforce Project*. 2. Be open and honest about needing to make any restructuring changes as part of the change your Salesforce project is bringing, and how you are helping affected people move on to other roles internally or outside your company. For the people you are enabling, training, and supporting to use Salesforce, emphasize that any restructuring that was planned to happen, is over.
Your change management initiatives related to management and organizational enablement are not executed or they are not executed effectively.	You critically need to have both group-level and local leadership on board and supporting the roll-out of your Salesforce solution. Their involvement should include the following: 1. Engaging with the immediate managers of your end users *before* training sessions and going live. 2. Leading the change by being present and opening training sessions to emphasize their support of the project and Salesforce solution. 3. Involving your local leadership teams with your project team, hypercare and support team, and with local change agents to understand their organization's adoption of your Salesforce solution and their sentiment towards it.

You have not involved users sufficiently in the development phase and UAT and as a result, you haven't ensured a great user experience that would enable them to become ambassadors for your project and Salesforce solution. This is, unfortunately, a common issue in Salesforce projects.	What you need to do is start involving and interacting with your users. Your approach can look similar to what's recommended in the next point – but rather than experiencing isolated pockets of resistance, you are likely seeing *widespread pockets of resistance*, and your efforts and approach should reflect that.
You have isolated pockets of users who complain and speak poorly of your solution, while a silent majority of users find it to be a good improvement for their work life.	Here, you should experiment with surgical interventions with the discouraged group of users. Have these for only one persona group of users to understand what their key experienced pain points are. It may be a lack of training in how to use Salesforce to their advantage. It may be they are fearful of any future restructuring. Whatever their main concern is, address them diligently – for example, with more training or reassurance that no more organizational changes are coming, and they can now focus on getting good at the new way of working and using Salesforce to do a better job. See how these interventions improve the overall organizational sentiment toward your solution to improve its adoption – make a note of the approaches and results in your change management plan template for further roll-out waves.
Some users don't use your Salesforce solution as intended. Instead, they may be finding shortcuts to "get their job done." This could be due to training issues or business processes not fully addressed in the design of your solution.	First, you should understand whether your solution fully supports the intended business process of your users. If that's not the case, you need a technical fix to the problem to mitigate the issue of "using the system wrongly" – and support your users to use it correctly for their benefit. If the system does support the intended processes of your users – meaning it's working as designed – you need to invest in further training of your users to support them in getting the most out of their new system.
Your data migration was not planned or executed well, leaving users with incomplete or disjointed historic Salesforce data.	Determine whether the mess with your data is due to poor data migration or process automations in Salesforce. To address this, you need to quickly gather information and analyze what data is affected and the root cause. If it is, in fact, due to poor migration of data from your legacy system(s), you need to execute data patches to improve the quality of your data.

Your data-reporting strategy was not prioritized sufficiently during the development phase, and as a result, your users and managers are left without coherent insights into operational performance for them to control their business.	This is an especially tricky one, as the successful enterprises of today and tomorrow are the ones that are able to leverage and act on the insights of the 360-degree view of their customers and of their business overall. We'll dive more into how you can make use of your Salesforce data in *Chapter 13, Managing Your Salesforce Data to Harvest the Fruits of Customer 360*. For now, you need to prioritize creating reports and dashboards for your organization to be able to manage your business. Consider the following approach: 1. Start by understanding the processes and KPIs of each business function or persona – this should be part of the original business case for your Salesforce project. 2. If Salesforce is the sole source of data to be able to track or calculate the KPIs – and the reporting requirements are simple – leverage standard Salesforce Reports and Dashboards. On the other hand, if some of the data needed to track and calculate the KPIs resides outside Salesforce – and the reporting requirements are complex – consider leveraging **Salesforce CRM Analytics** (which is part of the Salesforce platform) or utilize an external **business intelligence** (**BI**) platform (if already part of your enterprise system landscape) such as **Microsoft Power BI**, **IBM Cognos Analytics**, and **ClickSense**. You can check out the following link to review the Salesforce help article *Reports and Dashboards Limits, Limitations, and Allocations*: `https://help.salesforce.com/s/articleView?language=en_US&id=sf.rd_reports_dashboards_limits.htm&type=5`.
The majority of your users perceive and report the training delivered to be poor.	You should be asking for feedback after each training is delivered. Depending on the context and culture of the users being trained, this should give you some indication about the (perceived) quality of the training delivered- and make appropriate modifications to your training plan, content, and delivery. As a bare minimum, prepare materials for and hold direct user training (in-person or virtual), and critically, make the materials easily accessible for your users, and review and update them regularly.

You are unable to provide adequate hypercare because of the following:

1. The volume of tickets relative to the allocated team is too great, your hypercare team is spending more than 90% of their time on receiving bugs, processing them, and communicating back to the bug reporters – and you are not seeing a significant drop in this percentage. The consequence is your team does not have time to do proactive analysis and interventions to understand and reinforce user adoption.

2. Bugs and change requests aren't submitted in the right format, and your team is spending needless amounts of time gathering information to reproduce the bugs to find their root causes and fix them.

3. You transferred the responsibility of providing hypercare from the development team to your application management and support team shortly after going live, and there was no proper solution handover session to introduce the Salesforce solution to your team. This was done in an attempt to lower the cost of the implementation project. It is, however, a risky approach to hand over hypercare quickly. The development team members are the experts on your Salesforce solution, and, as such, are able to troubleshoot reported bugs much faster than any new support team member.

See the following mitigation or resolution for each point:

1. Although resources are scarce, you need to be bold and prioritize some of your team's time to investigate the root cause(s) of the volume and nature of the requests. It may be that a lot of "duplicate" bugs are reported. To mitigate this, make sure you systematically relate a reported bug to a user story. Attack the duplicate bugs in one go by one support team member. User training may have been inadequate and need to be refreshed – or the availability and location of training material need to be re-communicated. If you don't see a decrease in the toll hypercare is taking on your team, consider adding additional resources while continuing to perform root cause analysis, interview select users – and act on your findings to improve adoption.

2. Make sure the required bug reporting template is shared with all users and that the process for reporting them is communicated. If bugs are reported in other ways, kindly let the reporter know what the right template and process are.

3. The recommended approach is a soft handover from the development team to your support team – starting with user management and analyzing change requests. This will allow the support team members to gradually become familiar with your users, Salesforce solution, and setup. Analyzing change requests will make them familiar with the processes supported by your solution. Once they are up to speed with your solution, have them support in triaging reported bugs alongside your development team. Have your support team start building a **frequently asked questions** (**FAQ**) and answers knowledge base. Gradually assign an increasing number of bugs to your support team as their capacity allows.

You are overwhelmed by the required maintenance your solution requires – it may be for user provisioning for new employees, updating user permissions for employees changing positions, or for updating reference or master data. Any of these may lead your hypercare and ongoing support team to spend time on non-value-adding activities rather than driving adoption and supporting your users.	Depending on the nature of the required maintenance, you should consider the following steps: 1. Analyze whether investing in an identity and access management solution would make sense. This would alleviate much of the manual work for setting up new users and access changes to existing users changing internal jobs. 2. If the reference data or master data has its source of truth in another system, consider investing in system integration to alleviate the manual swivel-chair integration efforts your application management team is currently doing.
You are unable to get a bearing on the actual adoption of your solution since it varies by country, role, and persona. As a consequence, you struggle to prioritize who to interview to understand the reasons for low adoption.	Here, you need to do the following: 1. Recap the *Defining and measuring adoption* section in *Chapter 10, Communicating, Training, and Supporting to Drive Adoption*, to create great adoption and usage tracking dashboards for your project team and stakeholders. 2. Track your users' adoption – login and usage – of Salesforce and compare the adoption of users by persona group (business function) across your local organizations. Combine this data with the surveys you should have sent out to gauge your users' sentiment toward your Salesforce solution. Create a matrix containing four boxes: *Using regularly and liking it, Using regularly and not liking it, Not using regularly, and liking it*, and *Not using regularly, and not liking it*. 3. Tailor your approach to the user segments in the different boxes. Have conversations with ambassadors in the local organizations with higher adoption to understand what they particularly enjoy about your Salesforce solution – highlight these points when communicating with other user groups who struggle to see the upside. Prioritize interviews with the users who struggle most with adoption to understand and address their main pain points and concerns.

Critical bugs reported in production that *were or were not* present or discovered in SIT or UAT.	If the bugs *were not* present or discovered in SIT or UAT, fix the bugs, and regression test the business process. If the bugs *were* present or discovered, you need to do two things: 1. Fix the bugs and regression test the business process. 2. Investigate the root cause of how a bug managed to resurface, and then fix it.
Users are not able to distinguish between a bug and a new idea, and thus report new ideas as bugs.	Before deciding to "just fix the bug," analyze whether or not it is really a bug, an enhancement, or just a "perceived issue." You can do this by tracing the functionality being tested back to a user story or test case. It may be a bug, in which case your PO should prioritize the value of it being fixed against other backlog items. If not, it might just be an incorrectly written test case a change request, or a new feature request. Regardless of the outcome of this analysis, it is critical for adoption and user sentiment towards your Salesforce solution that you communicate back to the reporter what is being done (or not being done), and why. Consider having just one template for your users to report both bugs and new ideas.

Non-functional requirements (NFRs) not defined up front, and users report finding their Salesforce solution to be cumbersome and requiring more clicks compared to their legacy system. Users may also report performance issues, such as slow loading or updating of pages.	You can approach this issue in the following ways: 1. Define NFRs. If this is not possible for your stakeholders at this point, try to identify the key flows that are (reported to be) the most cumbersome in your Salesforce solution. Then, note down the number of clicks or the duration to complete the same business process in your legacy system. 2. Identify which *processes* in your Salesforce solution take (significantly) longer to complete compared to your legacy system. 3. Identify the *process steps* of the most cumbersome (new) processes that take the longest and find solutions to simplify or automate them. 4. Prioritize the potential changes by what will deliver the most improvement versus effort. 5. Communicate your effort and plan to your stakeholders at all levels. If users report performance issues, you need to identify the root cause for them. These could range from inefficient backend or frontend design to sub-optimal integration patterns or implementation. Regardless of the root causes, once they are identified, follow steps 4-5 to address the issues.
You have difficulties evaluating your original business case to know whether your project is a success – either because *no business case was created*, or it didn't include KPI targets as the basis for the business case.	If you didn't create a business case, you are really *in the blind* and unable to know whether your investment is paying off. What you can do is track how your business is performing some quarters after going live and compare it to business performance before going live. Be cautious of two things: 1. Don't equate correlation with causality – there may be any number of factors influencing your business performance, both positively and negatively. 2. Rolling out Salesforce won't (necessarily) result in instant improvement in business performance in the first week or month. It takes time for your users to get used to working in a new system. Remember, they may have spent years getting proficient in the "old" way of working, regardless of what system support they may or may not have had.

	To address this issue, you should do the following exercise:
The business case *was created*, but your business performance is varied, and you struggle to actually understand why.	1. Break the overall business case down by business function and geography in the scope of your Salesforce project. 2. Importantly, try to understand what the other factors may be impacting the business performance of the different business functions and geographies – which of the following may be the key drivers and root causes? Are some geographies not hitting the KPI targets? Are they hitting them, but not your adoption targets or vice versa? Is the overall market growing or slowing? 3. Create a grid to understand which countries/business functions are performing better than the targets in your original business case, and which are underperforming. 4. If there is correlation between business performance and the adoption and usage of your Salesforce solution, it is a prerequisite for proving causality. 5. Share your findings with the group and local leadership and chart an appropriate course of action to improve business performance.

Table 11.3 – Possible root causes for struggling with adoption and improving business performance

Next, let's evaluate your Salesforce project in the roll-out phase.

Evaluating the state of your Salesforce project

We have a lot of topics and activities in the roll-out phase of a Salesforce project. If you get stuck, or you are not pleased with the progress of your roll-out, this section will support you in evaluating the state of your Salesforce project. For each item you cannot check, go back to the corresponding chapters and complete the activities to proceed with your Salesforce project.

Checklist for your Salesforce project in the roll-out phase

In the roll-out phase, you should have gone through the following activities.

When deploying your Salesforce solution through environments, you should have done the following:

- Made sure you have perfected your deployment plan and checklist, also called a deployment runbook:

  - Critically including how you validate and sign off your deployments

- Performed SIT to identify and resolve merge conflicts with concurrent streams and/or what's already in production

Next, you should have completed these activities to perform UAT of your solution:

- Determined who should participate in and drive UAT

- Defined and aligned what process your acceptance testing team should follow and what template to use to report bugs and change requests

- Communicated back to bug and change request reporters the status and what you intend to do with them

Next, you should have completed these activities to check your data migration preparedness:

- Executed at least one successful dry run of data migration in a staging environment

- Ensured that the team performing the data migration is aware of their tasks and is booked for it – including your business and technical stakeholders who are to sign off the data migration

- Communicated the data migration plan and any expected downtime associated with it

- Have a fallback plan in case your migration fails or takes longer than accepted

Next, you should have planned, communicated, and managed the change to come by doing the following:

- Understood clearly what will change and for what stakeholders

- Created a phased roll-out plan if you are rolling out to multiple geographies

- Created a change management plan template and adapted it as you progress with your roll-out in waves, including for each roll-out wave:

  - Change readiness assessment

  - Communications plan

  - Overall roll-out plan

  - Management and organizational enablement plan

  - Change agent recruitment and enablement

  - Local technical deployment plan

  - Local data migration plan

Next, you should have done the following preparations to train your users:

- Determined your mix of training methods to maximize the impact

- Considered and planned for the training of your users as well as local leadership

- Communicated your training schedule and received confirmation for it from local organizations - and executed it

- Trained and enabled your application management, support team, data, BI, reporting, and analytics staff

Next, you should have done the following preparations to provide hypercare and manage production support:

- Determined your hypercare and ongoing support process

- Allocated resources and aligned roles and responsibilities

- Decided the criteria for transitioning from hypercare to ongoing support – hitting your adoption target

Last, but certainly not least, you should do the following activities to monitor and drive the adoption of your Salesforce solution:

- Created reports and dashboards to track login and usage of Salesforce

- Sent out and analyzed responses to surveys to understand your user's sentiment toward Salesforce

- Carried out interviews to better understand your user's perceptions and suggestions for improvement of your Salesforce solution

- Carried out interventions with individuals or groups of users resistant to change – to understand the underlying cause(s) and enable and support them to succeed with Salesforce

- Evaluated how your business is performing against the original business case for your Salesforce project – and saw it trending toward it

You may face a significant headwind in the roll-out phase of your Salesforce project. If you have been driving your Salesforce project by the book, you are remarkably close to the (immediate) finish line – a superior CRM solution for your organization to unlock greater business value.

Don't forget to pat yourself and your team on the back while continuing to experiment with change initiatives and interventions – and adapt your approach for future roll-out waves.

Summary

In this chapter, you have become familiar with the common issues often faced in the roll-out phase and learned how to mitigate or prevent them from arising. You have also assessed the state of your Salesforce project by going through the phase checklists.

With the roll-out of your Salesforce solution in great shape, you are ready to take your Salesforce program to the next level, the continuous improvement phase, starting with *Chapter 12*, *Evolving Your Salesforce Org and DevOps Capabilities*.

Part 4: The Continuous Improvement Phase

This part will teach you how to transform your organization into a product organization. We'll cover what you need to do to deliver **continuous improvement** (**CI**) for your users and organization to maximize the value derived from your Salesforce investments. We'll also discuss common issues so you can avoid or mitigate them.

This part has the following chapters:

- *Chapter 12, Evolving Your Salesforce Org and DevOps Capabilities*

- *Chapter 13, Managing Your Salesforce Data to Harvest the Fruits of Customer 360*

- *Chapter 14, Common Issues to Avoid in the Continuous Improvement Phase*

12
Evolving Your Salesforce Org and DevOps Capabilities

Following the roll-out of your Salesforce solution in the previous chapters, in this first chapter in *Part 4, The Continuous Improvement Phase*, of your Salesforce program, you will be guided through the nature of and activities in the continuous improvement phase. Then, we'll go through the concept of product organizations and why it matters for the success of your Salesforce program. Next, we'll cover program and platform governance in the continuous improvement phase. Finally, we'll introduce DevOps and how to evolve your Salesforce **DevOps** capabilities.

By the end of the chapter, you will understand the key activities in the continuous improvement phase. You will learn why you should aspire to become a product organization, and how you can start the **transformation** of your organization. After reading this and applying the concepts, you will be able to update and maintain your Salesforce roadmap and manage technical debt on your Salesforce platform. You'll be able to enhance your Salesforce DevOps capabilities to be successful in the continuous improvement phase of your Salesforce program.

In this chapter, we'll cover the following main topics:

- An overview of the continuous improvement phase
- Transforming your organization to become a product organization
- Governing your Salesforce program and platform
- Evolving your Salesforce DevOps capabilities

Let's go!

An overview of the continuous improvement phase

This phase of the Salesforce program begins the moment you go live with your first users, as illustrated in *Figure 12.1*, and it doesn't end (unless your organization decides to stop using Salesforce).

Figure 12.1 – Continuous improvement concurrent with roll-out

We can summarize the key *continuous* activities as follows:

- Transforming your organization to become a product organization
- Governing your Salesforce program and platform
- Evolving your Salesforce DevOps capabilities
- Managing and governing the quality of your data
- Leveraging your C360 data to maximize your Salesforce investments
- Leading and inspiring the peers in your organization

The last three topics will be covered in the next chapter.

Let's get started with product organizations.

Transforming your organization to become a product organization

In the last decade, digital transformation initiatives – recently accelerated by the COVID-19 pandemic – have become ubiquitous throughout medium and large organizations. Literature is quick to point to the failure rate of these initiatives – the reasons cited often being a lack of user involvement, cross-functional coordination, executive commitment, and strategic alignment. Most of these areas are addressed by the **product organization**.

The concept of **product management** originated from Procter & Gamble, where Neil McElroy wrote a memo requesting to hire people and structure them around a single product or brand. The ultimate responsibility of *brand men* would be to manage the product life cycle, sales, and promotions – supported by field testing and customer interaction. Many additional and complementary concepts, theories, and practices have emerged since then, including Kanban (1940s), Kaizen (post-World War II), The Agile Manifesto (2001), DevOps (2007), and Lean Startup (2011).

> Tip
>
> If the concepts of digital product management and product organizations are entirely new to you, you can start your learning journey by reading *The Lean Startup* by Eric Ries and *Inspired: How to Create Tech Products Customers Love* by Marty Cagan.

That's all great, but how does it all relate to a modern, tech product organization? Let's start with some definitions.

> *A product organization is an organization that is constantly trying to improve its product offerings, which are comprised of features to provide benefits to its customers.*

Let's translate this to a Salesforce program context:

> *It is an organization that is constantly trying to improve its Salesforce solutions to provide value to its Salesforce users (employees, customers, and partners).*

According to the **agile maturity model** in *Open Agile Architecture* – a standard of **the Open Group**, which you can check out at `https://pubs.opengroup.org/architecture/o-aa-standard-single/` – an organization cannot progress to the higher levels of agile maturity without adopting the traits and practices of product-led organizations.

> Important note
>
> While a lot of literature primarily focuses on building great products for *external* customers, the concept and practices of a product organization can easily be translated to building Salesforce products for your external customers and partners, as well as all your *internal* users.

Let's start by identifying your Salesforce platform's products.

Understanding your products

Digital products are *vehicles to deliver value*. That may seem vague and abstract, which is why many organizations struggle to define their products, as illustrated by *Figure 12.2*:

Figure 12.2 – Comic Agilé #218 – the product organization

Jokes aside, what specifically your organization's *Salesforce products* are depends on the nature of your organization, users, customers, and partners.

Let's see an example of *products* built on the Salesforce *platform* for PME's *customers*:

- A Salesforce *app* for PME's *sales reps*

- A Salesforce *app* for PME's *customer service reps*

- A **Commerce Cloud** *storefront* with self-service functionality for PME's *customers* to place orders and log cases

- A Salesforce **Experience Cloud** *site* for *distributors* to collaborate with PME on leads and opportunities

Next, let's get aligned.

Aligning the vision, strategy, and ownership of your Salesforce platform

The good news is that if the development and roll-out of your initial Salesforce release have been similar to what is described in the previous sections of this book, then you have already *done* parts of what it takes to *act* as a product organization. You assigned and enabled a **product owner** to own the product you offer to your users.

Now, *doing* and *acting* as a product organization is not the same as *being* a product organization – the transformation spans multiple dimensions, including the following key ingredients:

- Deep and continuously growing *insights* into your Salesforce users' pain points and needs, and what products (Salesforce solutions) you are offering them to make their life easier and more productive
- *Organizational alignment* to transform from a project-based organization to a product organization, which includes the following:
  - Aligning your business strategy to your Salesforce platform and product strategy.
  - Transforming your funding model *from* funding projects (statements of work) *to* funding **cross-functional product teams** that continuously deliver incremental value. This critically includes tying funding to a product owner mandate to prioritize roadmap items – derived from a fine balance between user insights and the strategic intent of your organization.
  - Stakeholder buy-in and involvement in your Salesforce product roadmap.
- Maturing your Salesforce platform governance structure, driven by your Salesforce **Center of Excellence (CoE)**
- Redefining your (IT) **operating model** and enhancing your Salesforce DevOps capabilities

With your platform and product vision, strategy, and ownership aligned, let's set some goals.

Setting goals for your Salesforce platform and products

Salesforce is a broad suite of SaaS clouds and add-ons, as well as a PaaS. This means you can extend and enhance the **out-of-the-box (OOTB)** functionality and apps – and build your own.

For most organizations, investing in Salesforce is a strategic decision to unify and consolidate into one CRM platform. Their vision is rarely for their Salesforce platform to serve just a single user group with one solution (cloud, product, or app); rather, it is to be a **strategic platform** with multiple products for multiple user groups.

If that's also the goal for your organization, this means you need to consider and align your vision and strategy at both the Salesforce *platform* and *product* levels.

> **Tip**
>
> If your organization's vision and strategy aren't crystal clear for you, engage with the sponsor of your Salesforce program. You can also recap *Chapter 1, Creating a Vision for Your Salesforce Project*, for methods and inspiration to go about determining a vision for your Salesforce program and platform.

With your Salesforce platform and product strategy aligned, you need to set **product goals** for your Salesforce platform and products.

If you have multiple products with different product owners, you should define product goals at the Salesforce platform level and for each product domain – for example, sales, customer service, field service, commerce, and digital marketing.

> **Important note**
>
> **Organizing in cross-functional agile teams** – an agile team needs to have one, and only one, PO, and one agile team only has one backlog. A developer (functional or programmatic) should only be a member of one agile team (scrum masters and QA specialists can potentially serve up to two teams). The reason is simple; if a developer is part of multiple agile teams – with different POs – the question arises of how the developer should decide which product backlog to work on. Solving that puzzle would require inefficient inter-team planning between scrum masters to determine capacity allocation for each team, not to mention the inherent inefficiency caused by the daily context-switching of the developer.

Platform and product goals serve the purpose of helping guide your teams to prioritize what to build next.

> **Important note**
>
> Although, in this section, we have been discussing organizational transformation to become a product organization, you can – if the concept of a product organization is completely new to your organization – opt to start with your Salesforce program and platform to lead the way for your organization.
>
> Beware of potential resistance, and spend a lot of time engaging, discussing, and sharing the concepts with your stakeholders to get them on board. Your Salesforce program sponsor needs to be intimately involved in the process.

In this section, you have learned about product organizations, how to go about transforming to become one, and how to align your organization's strategy to your Salesforce platform strategy and vision. Let's cover governance in the continuous improvement phase and how creating a roadmap ties into your strategy and product goals.

Governing your Salesforce program and platform

With many changes happening to your Salesforce program and the platform to serve your organization, having solid governance in place is key for stability and growth.

Let's begin by looking at your Salesforce CoE role in the governance of your Salesforce program and platform.

Updating your Salesforce CoE

Your Salesforce CoE is still the overarching governance body responsible for overseeing your Salesforce program. It consolidates knowledge, guidelines for ways of working, and change management, and manages the governance structures and forums for business, technical, and data governance.

The key leadership roles in your Salesforce CoE in the continuous improvement phase are as follows:

- **Salesforce platform owner**: This is the owner and overall accountable for your Salesforce platform, and its roadmap:

  - Some organizations also refer to this role as **chief product owner** or **product manager**. If your Salesforce platform offers multiple distinct solutions to different user groups, you may have different product owners of those products:

    - Although you may have multiple products, you can group them by domain or capability to simplify ownership and product management structure.

> **Important note**
>
> It is critical that you tie the program, platform, and product funding mandate directly to the platform and product owner(s) to empower them and materialize ownership.
>
> You may think such a mandate is only to be given to senior leaders within your organizations – and you would be absolutely right. The Salesforce platform and product owner are indeed senior positions suitable for a senior, experienced professional.

- **Salesforce platform architect**: They are responsible for the architecture vision of your Salesforce *platform*, as well as your Salesforce organization's performance, reliability, security, and flexibility.
- **Salesforce program owner**: They are responsible for organizing and structuring your Salesforce program, resourcing and budgeting, and facilitating governance forums.
- **Salesforce program sponsor**: This is the executive sponsor from your organization's senior leadership who provides guidance and support.

Revisit *Chapter 3, Determining How to Deliver Your Salesforce Project*, for foundational CoE setup.

Managing, governing, and leveraging your Salesforce data will be covered in *Chapter 13, Managing Your Salesforce Data to Harvest the Fruits of Customer 360*.

Let's look at your partner model.

Adjusting your partner model in the continuous improvement phase

As you transition from the development and roll-out of your initial Salesforce release to the continuous improvement phase, there are a number of reasons why you should consider modifying your partner model:

- **Capabilities**: The implementation partner of your initial release was qualified and capable to support you in the implementation of the technical scope of your MVP but does not cover the upcoming capabilities on your Salesforce roadmap:

  - This makes sense, as most implementation partners are capable of supporting the implementation of core Salesforce clouds such as Sales Cloud and Service Cloud, but not all are experts in products such as **Commerce Cloud**, **Mulesoft**, **Revenue Cloud (CPQ)**, **Field Service Lightning**, and **Marketing Cloud**

- **In-house strategy**: It is common for organizations to have an in-house team to focus on Salesforce administration, user management, ongoing maintenance, support, and the development of change and new feature requests to *existing* currently supported capabilities:

  - Few organizations have the scale for their in-house team to be able to cover all the preceding items, in addition to having the business and technical skills to be able to drive large initiatives, such as extending with new clouds for new user groups

Next, consider which of the following areas you think is the most appropriate to be managed by your internal team, a partner, or a combination:

- Basic user management and simple help desk tickets
- Managing small change requests
- System monitoring and evaluating Salesforce's three releases per year
- Agile product teams delivering new features
- Technical design governance
- Program management and mid-term planning
- Change management and communication

When determining this, consider whether or not each area is within your organization's core competencies. If an area is not within a core competency, consider leveraging the expertise of a partner.

With your Salesforce CoE and partner model squared away, let's move on to your Salesforce roadmap.

Revisiting and maintaining your Salesforce roadmap

Recall the roadmap you created in *Chapter 2, Defining the Nature of Your Salesforce Project*. The business owner in the pre-development phase of your Salesforce journey may have become the product owner in the development phase of your initial Salesforce implementation.

What should they do after your initial release is developed and being rolled out? Go back to their old job? No! There is a great (and continuous) job to do to further improve and maximize value from your Salesforce platform and products for your users and organization.

Here are some of the sources for the backlog items your platform owner and product owner(s) need to comprehend, consolidate, and ultimately prioritize:

- **At the beginning of the continuous improvement phase of your Salesforce program** (shortly after the initial go-live of your first release), do the following:

  - Review the capabilities your organization originally considered for your Salesforce project:

    - If you have phased your Salesforce roll-out by scope – for example, by business domain (such as sales, service, and marketing) – you will have these as input for your roadmap

  - Items that were de-scoped in different phases of your initial implementation:

    - De-scoped epics and user stories in the upfront architecture phase

    - Feature or change requests captured in sprint demos and UAT in the development phase

    - Feature or change requests captured during training, and during hypercare of the roll-out phase

- **Continuously improve your Salesforce products**: The sources of input for the continuous improvement of your Salesforce platform products include the following:

  - Change requests from users

  - New feature requests from users

  - Changing strategy and priorities

  - Changing market conditions

  - Organizational changes

  - Technical innovation – with three releases per year for most Salesforce products, there are plenty of new features to consider taking advantage of (if they would add value to your users and organization); after all, you are paying for them

What most organizations struggle with is translating new medium-to-large feature requests into new capabilities (epics) for their existing Salesforce *products*. This makes great sense – your organization is paying for licenses that include many OOTB functionalities that could be configured and customized to make work easier and more productive for your users. Moreover, as your users start to use Salesforce, they become invested in making it work to their advantage and, therefore, request more features.

Make sure you have a structured process for listening, collecting, and evaluating feedback, as well as for communicating your decisions.

> **Tip**
> Consider using a **product management system** that serves as the dedicated system of record for features for product managers (Salesforce platform and product owners) and helps them create and maintain roadmaps.

Let's look at how you can distinguish the nature of different product backlog item types:

Type	Product backlog item example	Decision-making model	Lead-time target
Theme	Requests for new Salesforce *products* (for example, new Salesforce clouds such as **Service Cloud**, **Marketing Cloud**, and **Commerce Cloud**, or major add-ons such as **CPQ** and **Field Service Lightning**) to support new user groups on your Salesforce platform. These may be original roadmap ideas you have decided to phase for later implementation or new requests based on innovation and Salesforce's offering.	Similar to the process described in *Part 1, The Pre-Development Phase*, of this book	One or more program increments. Alternatively, follow a hybrid delivery methodology to build the first release of the new solution.
Epic	New, larger feature requests to support new capabilities on *currently used clouds*. The original roadmap epic, phased for later implementation on existing clouds.	Salesforce PO's epic prioritization shared with the **program board**, based on benefit hypothesis analysis	One or more sprints, but less than one program increment.

Type	Product backlog item example	Decision-making model	Lead-time target
User story	Change requests and enhancement requests to existing features on *currently supported capabilities*. Also created for new epics.	PO determined by ordering their **product backlog**, based on business value and alignment with product goals.	Must be able to complete within one sprint. If not, break into more user stories.

Table 12.1 – Salesforce product backlog item types

As described in *Table 12.1*, adding an entirely new theme (a new Salesforce cloud or major add-on) is a large undertaking that you should consider approaching in much the same way as described in previous parts of this book.

> **Tip**
>
> When you evaluate requests for new features and enhancements, check out the Salesforce Roadmap Explorer at `https://architect.salesforce.com/roadmaps/roadmap-explorer` to see whether any of the upcoming releases may be able to support the request.
>
> It is usually better to plan for the adoption of new OOTB features rather than building your own. Be aware that the roadmap explorer describes intentions, not a commitment – make your decisions accordingly.

You need a structure and process for continuously collecting, qualifying, and prioritizing new requests. SAFe offers different configurations to manage new ideas, such as the **program board** with a **program KanBan**, and regular cadence to review, provide feedback, and support the progression of epics to the next stage. A PO should be able to quickly explain their decisions and prioritizations, as it should be immediately evident how an epic's benefit hypothesis ties back to the product goals. As such, the program board is not intended to be a decision-making body, nor should it be a forum for long-winded discussions and presentations.

> **Tip**
>
> Your PO can, in addition to benefit hypothesis and an epic's alignment to product goals, use the **weighted shortest job first** (**WSJF**) model, as described in the SAFe framework. You can read more about it here: `https://www.scaledagileframework.com/wsjf/`.

You should determine how to progress your roadmap epics to the development pipeline. There are different options for translating your Salesforce roadmap to mid-term plans. Depending on the scale of your Salesforce program and the number of agile teams, you may opt for a smaller or larger planning setup.

Two common frameworks for large agile setups are **Large-Scale Scrum (LeSS)** (`https://less.works/`) and SAFe's **program increment (PI) planning** (`https://www.scaledagileframework.com/pi-planning/`).

> **Important note**
>
> You should choose a framework that fits *your* organization's needs, context, and level of agile maturity to deliver continuous improvement.

Let's now turn to the illusive concept of technical debt.

Managing technical debt

Technical debt is best described by its symptoms and how it is created.

The symptoms – or the impact – of technical debt include degrading system usability, slowing performance, corrupted data, and increasing security risks and issues. These symptoms materialize as an increasing number of bug reports, decreasing adoption rates, and a slowing innovation rate.

Technical debt is created – or accumulated – in the following ways:

- Making poor changes to your Salesforce platform based on the following:

  - *Poorly designed* solutions that require a high degree of maintenance outweighing the benefits of the solution itself

  - *Poorly governed* solutions that have been extended beyond the scope of their original purpose

- Creating great solutions to the *wrong problems*:

  - If a solution is designed well but does not address the right problem, it won't deliver business value, only clutter – technical debt

- *Not making any changes* to your Salesforce platform:

  - The world is changing at an accelerating speed, and organizations change to stay ahead or keep up with those changes. If the products – your Salesforce solutions – you offer to your users don't change with your organization, they will effectively eventually become technical debt.

It's impossible not to have any technical debt, and it's not a one-off activity to remove the debt you do have; rather, it's a continuous effort to manage it. That doesn't mean you should do nothing – on the contrary!

You need to implement *governance structures* to limit the amount of new technical debt introduced to your Salesforce platform and *continuously remediate* impeding debt from your organization.

> **Important note**
>
> Make sure your Salesforce platform architect works closely together with your platform owner and product owner(s) to ensure architecture principles and technical enablers (refer to *Chapter 2, Defining the Nature of Your Salesforce Project*) are included for changes and new solutions for your Salesforce platform.

We'll address the common issues and strategies for preventing and mitigating technical debt in *Chapter 14, Common Issues to Avoid in the Continuous Improvement Phase*.

Having established your Salesforce roadmap – balancing functional features, technical enablers, and refactoring to manage your technical debt – let's now turn to your Salesforce DevOps capabilities.

Evolving your Salesforce DevOps capabilities

To unlock more value from your Salesforce investment and achieve success while having peace of mind in the continuous improvement phase of your Salesforce program, you need to understand and enhance your organization's DevOps capabilities.

The process and capabilities you used in the development and roll-out phase of your initial Salesforce release were suitable for those phases – initial implementation and roll-out. *Now*, you need to set up your Salesforce program to be able to *continuously* take in feedback, build, release, and ensure the usage and adoption of new features – all while making sure your Salesforce organization is performant, reliable, and secure.

Let's start by introducing DevOps.

Introducing DevOps

In 2004, Charles A. O'Reilly III and Michael L. Tushman published a paper titled *The Ambidextrous Organization* (you can read the paper here: `https://hbr.org/2004/04/the-ambidextrous-organization`).

What characterizes an **ambidextrous organization** is its ability to do two distinct things well, *simultaneously*. What are those two abilities then? The ability to *manage operations* and the ability to *develop the organization*. Sound familiar? Yes, it's awfully similar to DevOps – a movement that didn't really begin until a few years later in 2007–2008.

The paper proved ambidextrous organizations were top performers among their industry peers and 90% of them consistently reached the goals they set out to achieve.

Not surprisingly, I see the organizations that are most successful in implementing and evolving Salesforce are the ones that are able to balance creating a usable, performant, reliable, and secure Salesforce platform while continuously delivering incremental business value to their users. It's a case of staying grounded while reaching for the stars.

While the theory of ambidextrous organizations is a general theory for an entire enterprise, DevOps was, and is, focused on the domains of development and operations of IT systems.

Understanding the principles and goals of DevOps

As the use of agile delivery methodologies has become more prevalent since emerging in the 2000s and 2010s, the speed at which development teams are able to deliver new features has accelerated. As great as that sounds, it has, however, to some degree, come at the expense of decreasing system reliability. This means both a more unstable user experience as well as a headache for IT ops teams.

The DevOps movement emerged to address those pain points by bringing together the IT development and operations teams, as well as *the business*, to further a *holistic mindset with shared goals* of how value is delivered while maintaining stable operations, as represented by the continuous DevOps loop shown in *Figure 12.3*:

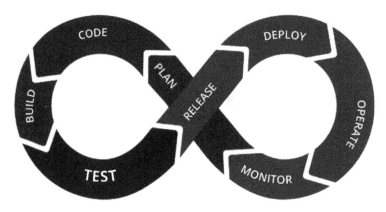

Figure 12.3 – The eight steps of the continuous DevOps loop

According to the 2022 **DevOps Research and Assessment** (**DORA**) *State of DevOps*, there are five metrics to measure an organization's DevOps capabilities:

- One metric is related to *operational performance*:

  - **Reliability**: This covers availability, latency, performance, and scalability

- Four of the metrics are related to *software delivery performance*:

  - **Lead time for changes**: The time from when changes are committed until they are released to production

  - **Deployment frequency**: How often you deploy to production

  - **Change failure rate**: The percentage of deployments that either fail or require hot-fixing or another rework to make them work

- **Mean time to restore**: This is related to the previous metric – measuring the time it takes to restore a service after a change failure or incident

> **Tip**
> You can read the full 2022 DORA State DevOps report here: `https://services.google.com/fh/files/misc/2022_state_of_devops_report.pdf`.

Let's now turn to Salesforce.

Enhancing your Salesforce DevOps capabilities

So, what does DevOps mean in a Salesforce platform context? It is optimizing how value is delivered to the users of your Salesforce platform and products.

So, what's different in the continuous improvement phase? In the hybrid agile model, the epics and user stories were defined upfront before starting development, and they went through development and testing before a dedicated phase for SIT, UAT, production release, and user onboarding in the roll-out phase. Now, in the continuous improvement phase, you should aim to adopt DevOps to be able to execute all of those steps and frequently release changes.

Let's break down each of the eight steps of the DevOps process (refer to *Figure 12.3*) and see what it means in a Salesforce context, and what tools your organization can consider to support it.

Streamlining your Salesforce operations

While DevOps absolutely is an integrated process, there are parts that are predominantly ops focused, and some that are dev focused.

We'll start with Salesforce *ops*, also known as **Salesforce administration**, for the simple reason that if something is broken, you can't effectively develop it.

The goal of Salesforce administration is to ensure reliability, performance, and security.

Core Salesforce administration entails the following:

- Managing your Salesforce production environment and providing support for your users
- Potentially managing a production support environment
- User management:
  - Provisioning and de-provisioning new joiners and leavers
  - Employees change roles and need new permission sets, groups, and queue memberships

> **Tip**
>
> Read the Salesforce Admin blog post about the future of user management at `https://admin.salesforce.com/blog/2022/the-future-of-user-management`.

Depending on an organization's scale, Salesforce administration may also entail the following:

- Managing, maintaining, and creating sandboxes – including data
- Handling new feature requests from users
- Handling bug reports from users
- Maintaining training materials
- The first line of support for questions following the training of new users/employees
- Regression-testing key processes three times a year with every Salesforce release
- The support and creation of reports and dashboards
- Managing data and performing data load requests (we'll cover data management and governance more in *Chapter 13, Managing Your Salesforce Data to Harvest the Fruits of Customer 360*)

> **Tip**
>
> While ensuring a stable, secure, and reliable Salesforce organization for your users is essential, be careful not to trap yourself by becoming too rigid, super-optimizing admin processes, or assigning too many resources for *operations* at the expense of resources for *development*.

Another key aspect of operations, which should be intertwined with Salesforce development, is **monitoring**. The goal of monitoring is to observe adoption, system performance, reliability, and security. Critically, the goal is also to *continuously collect feedback for improvement*.

The following activities describe the ways you should be monitoring your Salesforce org:

- Tracking and understanding the adoption and usage of your Salesforce products
- Monitoring platform performance and security

> **Tip**
>
> Check out the SalesforceBen article for free ways to monitor your Salesforce organization at `https://www.salesforceben.com/free-ways-to-monitor-your-salesforce-org/`.

- Performing regular technical and security audits
- Monitoring incoming feature requests

- Monitoring incoming bugs

- Facilitating sprint retrospectives and demos

> **Tip**
>
> Here are a few ideas you can use to improve how you collect feedback:
>
> • Facilitate user feedback forums.
>
> > • Regular cadence with superuser representatives from the users of your Salesforce products. Host forums per Salesforce product and user group.
>
> • You should provide a simple template and method for your users and other stakeholders to submit new feature requests and changes to existing ones.
>
> • Use a **product management system** to consolidate user feedback, validate ideas, create roadmaps, and integrate with your development tool. Leading product management system vendors include **Productboard**, **Aha!**, and **JIRA Discovery**.
>
> • While development-focused tools such as Jira and Azure DevOps have some of the same features as a dedicated product management system, their focus – and specialism – is to support the agile development process, *not* user insights collection, ideas validation, prioritization, and Salesforce product roadmap creation.

Let's move on to the development of new features for your Salesforce platform users.

Optimizing your Salesforce development process

The Salesforce development process starts with **planning**. The goal of planning is to order the product backlog and get the user stories towards the top of the product backlog to meet the **definition of ready**.

The definition of ready is unique to each organization. Let's see how our scenario company, **Packt Manufacturing Equipment** (**PME**), defined their definition of ready:

- The user story title, need, purpose, context, and details

- The epic it is related to

- Acceptance criteria

- Business value

- The solution design

> **Tip**
>
> You should also consider requiring user stories to state any required maintenance of the solutions proposed.

- **Change impact**, assessed for the following:

  - **User impact**: How big a change does this user story provide for users?

  - **Technical impact**: How much does the user story impact your technical architecture?

  - **Regression test impact**: How much does the user story change and interfere with the business process in context? Technically, how many metadata components does the user story interfere with or change?

> **Tip**
>
> There are many tools available to help you analyze the technical and potential regression test impact, including **Elements Cloud**, **Metazoa**, **Salto**, and **Happy Soup**.

- If the nature* of the user story requires it, its solution design is reviewed by the **design authority (DA)**

> ***Tip**
>
> You can opt to only require the DA to review user stories where the solutions contain changes that potentially pose a greater technical and security risk to your Salesforce platform. These can include data model changes, security and sharing model changes, integrations, programmatic development, and complex automations. You can read more about setting up a design authority in the *Governing your Salesforce project in the development phase* section in *Chapter 7, Building and Testing Your Initial Release.*

Why is it important for user stories to meet the definition of ready before starting development? Because you need to know what is going to be developed and released is based on a clear understanding of user needs. Moreover, that it is designed in line with both your development guidelines and best practices, and in a way that ensures users will want to use it.

You may decide to hold regular **backlog refinement** sessions to support your product owner in creating and qualifying product backlog items and to order their product backlog. In a backlog refinement session, a product owner, a business analyst, a customer experience expert, and perhaps a solution architect will typically do the following:

- Create and qualify product backlog items

- Assess change requests and user feedback forum input

- Review incoming bugs

- Break new epics into user stories

- Assign stories for solutioning ahead of design authority sessions

- Order the backlog ahead of sprint planning

> **Tip**
>
> Your product owner can use the following methods and measures to qualify user stories and determine their order in the product backlog:
>
> • How each item ties back to the product goals
>
> • The business value it delivers

If you are using Scrum, **sprint planning** will happen on a regular cadence (every 1–4 weeks), with the goal of determining a sprint goal and defining the sprint backlog by filling it up with user stories to the capacity of the agile development team. You may recall the options for defining and measuring sprint capacity in *Chapter 7, Building and Testing Your Initial Release*. To progress your agile maturity, you should consider using the pure agile concept of story points.

Sprint planning goes as follows:

1. Before sprint planning, the PO orders the product backlog.

2. The PO presents the user story at the top of the product backlog.

3. Story points are assigned using **scrum pokering**.

> **Tip**
>
> For virtual story point pokering, `https://www.scrumpoker-online.org` is a great online tool.

4. The user story is assigned to the sprint backlog by the scrum master.

5. The process is repeated until team capacity (story points) are assigned.

> **Tip**
>
> The first sprint planning when you introduce story point pokering will be an abstract exercise, as your team members won't (yet) have a common, shared frame of reference of what the nature of a seven-story-point user story is. Neither does your team have a history of sprints to know what their capacity is. The important thing is just to get started – take a leap of faith. The common frame of reference is typically established quite quickly, within one or two sprint planning sessions. For development capacity, you need to go with a vote of confidence from the team members in the first sprint planning. Experienced developers have a good gut feeling of how big, complex, and risky different user stories are, and they will share when they are losing confidence in the team's ability to complete the sprint backlog.

Next is **coding** or the actual development done by functional consultants and developers. The goal with development – not surprisingly – is to configure and develop solutions on your Salesforce platform.

Refer to the *Setting up your development model and environments* section in *Chapter 7, Building and Testing Your Initial Release*.

After development comes **building** or **continuous integration**, which entails merging the newly created or changed metadata with the existing version of your Salesforce source, and identifying any merge conflicts and resolving them. Refer to the *Deploying your Salesforce solution through environments* section in *Chapter 9, Deploying Your Release and Migrating Data to Production*, to read more about **system integration testing** (SIT).

When choosing your deployment (CI/CD) setup and tooling, you face the same challenge as when deciding how to build your Salesforce features – use OOTB Salesforce, buy and use an AppExchange solution, or build your own solution. Let's understand the options as of January 2023:

- **Salesforce OOTB deployment tools**: Salesforce's own tools include the following:

    - **Change Sets**: A legacy deployment tool. It's cumbersome, inflexible, and doesn't rely on a **version control system** (VCS).

    - **DevOps Center**: This recently released feature (GA in winter 2023) has some capabilities somewhat similar to the **COTS** (**Commercial Off The Shelf**) tools described further on:

        - It currently does not include features such as impact analysis, data migration, and choosing delta or full deployment, and can't be extended in the same way you potentially could if you built your own or were supported by some of the COTS CI/CD vendors.

- **Buying a COTS CI/CD solution**: Investigate which one of the many vendor solutions is best for your organization. Popular vendors include the following:

    - **Gearset**, **Copado**, **Salto**, **Metazoa**, and **AutoRABIT**

- **Building your own CI/CD pipeline**: This provides full flexibility but requires a dedicated team to manage and improve the setup.

> **Important note**
>
> Engage a Salesforce technical architect and/or a DevOps architect to analyze which DevOps tooling and roadmap is best suited for your organization, keeping in mind your resources, capabilities, and the scale of your Salesforce platform.
>
> Most of the COTS CI/CD tools mentioned leverage a Git-based VCS. Setting up a VCS is one of the basic prerequisite steps in implementing DevOps. Setting up and working with a VCS for Salesforce can be a challenge for non-technical team members. As such, be mindful of your intended team members when selecting a tool – and be sure to allocate time for training on how to use it.
>
> To help you choose a course of action, read the blog post *Which Git hosting provider is right for your Salesforce team?* from Gearset at `https://gearset.com/blog/which-git-hosting-provider/`.

Following a successful build or integration of components, the next step is **testing**. The goal of testing is three-fold:

- Testing the individual functionality of the user story:

  - This includes positive and negative test cases

- Regression-testing the process that the user story is part of to check that it works as intended
- Unit testing

When considering implementing an automated testing tool, be aware that regardless of how you execute regression testing, the tests need to be regularly updated as new features are added to your Salesforce organization, and this needs to be considered as part of the investment. Options to consider include **Provar**, **Testim**, and **Selenium**.

The next step is **deployment** to production, which is getting your merged, integrated, and tested solution(s) deployed to your Salesforce production environment. The goal of deployment in DevOps is continuous deployment, meaning that merged and tested solutions should be deployed as often as fast as possible to minimize the risk of them becoming outdated.

Finally, there is the matter of **releasing** your solutions to users.

Why are deployment and release separated? They aren't always, but it makes sense from a conceptual standpoint. Simply deploying a solution does not ensure users will use them. The goal of releasing is to be able to **release on demand** (when you and your users are ready) and ensure usage and adoption.

In Salesforce, **Custom Settings**, **Permission Sets**, and **Custom Permissions** are common ways to implement **feature flags** to release on demand.

> Tip
> To learn more about using Custom Permissions for feature flagging, check out this blog post at `https://blog.texei.com/why-we-should-use-custom-permissions-a8b5c22bbe94`.

To make sure users will use (and want to use) your new features, you need to help, guide, and support them – change management doesn't stop in the continuous improvement phase of your Salesforce program. Your change management effort should correspond to the scale of change.

Depending on the nature and scale of the change, your change and communication efforts could include the following:

- Release notes
- Training
- Videos

> **Tip**
>
> To enhance user adoption and minimize the need for synchronous training, consider using Salesforce **In-App Guidance** or, for more sophistication, AppExchange solutions such as **Improved Help**.

And with that, you have completed the loop from operations to releasing. Let's now reflect on and assess your current Salesforce DevOps setup.

Reflecting on your Salesforce DevOps journey

Your DevOps systems, as well as other sources, will tell you how well your DevOps capability is performing to deliver a stable operating environment, while continuously innovating for the benefit of your users and organization. It's a never-ending journey to refine your DevOps capabilities.

> **Tip**
>
> Take the **DORA DevOps Quickcheck Quiz** at `https://www.devops-research.com/quickcheck.html` to see how your organization's Salesforce DevOps setup compares to your industry peers.

Although the hardcore DevOps metrics will tell you how well your DevOps setup is performing, be sure to also measure how satisfied your Salesforce team members are with their work life. Job satisfaction of your team is inherently important to strive for, and also has a huge impact on the success of your Salesforce program.

> **Important note**
>
> Adopting a DevOps mindset and practices will require redefining your (IT) operating model and how your IT development department, operations department, and business stakeholders work together, and how they are organized.

Let's now wrap up this chapter!

Summary

We have covered a lot of topics in this first chapter in *Part 4, The Continuous Improvement Phase*, of your Salesforce program. First, you learned about the nature of activities in the continuous improvement phase. Then, we introduced the concept of product organizations, why you should aspire to become one, and how you can start the transformation of your organization. Next, we discussed program and platform governance in the continuous improvement phase. Finally, we talked about the history of DevOps and how to evolve your Salesforce DevOps capabilities.

You have now transitioned well into the continuous improvement phase of your Salesforce *program* – beyond initial *project* implementation. As you implement the concepts and structures laid out in this chapter, you should begin to see an increase in the rate at which value is delivered to your users and a more reliable system.

In the next chapter, we'll cover governing and leveraging your **Customer360 data** to unlock the full potential of your Salesforce investment.

13

Managing Your Salesforce Data to Harvest the Fruits of Customer 360

In this chapter of *Part 4, The Continuous Improvement Phase,* of your Salesforce program, we will cover how to manage your Salesforce data and how to leverage it to maximize the value of your Salesforce investments. First, we'll introduce the life cycle of your Salesforce data and how to govern its quality, access, and usage. Next, we'll explore the many use cases for leveraging your Salesforce data across different domains, using different Salesforce features and technologies to provide better user and customer experiences. Finally, we will show how you can inspire peers in your organization by showing the value of a successfully implemented, run, and leveraged Salesforce program and platform.

You will learn what it means to manage the life cycle of your Salesforce data in order to optimize the user and customer experience and the performance of your Salesforce org and ensure **business continuity**. You will learn about the **attributes of data quality** and how to improve the quality of your Salesforce data. You will learn how your Salesforce **Center of Excellence (CoE)** can structure its **data governance** efforts by establishing and running a **data governance board** – when your data and its quality is managed and governed well, you can begin to reap the fruits of your **Customer 360** – the view (and data) of your customers. You will learn about the ways in which you can leverage the **Customer 360** data at your disposal to automate processes and create superior, personalized experiences that drive desirable business outcomes – unlocking the full value of your Salesforce investment. You will also learn how to **inspire** and **lead** your organization on its wider journey of digital transformation.

In this chapter, we'll cover the following main topics:

- Managing and governing your Salesforce data
- Leveraging your C360 data to maximize the return on your Salesforce investments
- Leading and inspiring the peers in your organization

Let's go!

Managing and governing your Salesforce data

Data is the business currency of the present and the future. Some might even argue it is your organization's most valuable asset, other than your employees, of course.

Important as it is, many organizations struggle to truly leverage the data they have at their disposal. Next to siloed data, I would argue that data management and the **governance** of the quality of their data is the biggest challenge preventing organizations from reaching their full potential with Salesforce.

Let's jump right in and look at your data life cycle.

Managing the life cycle of your Salesforce data

To be able to effectively manage and govern your Salesforce data, we first need to understand a few definitions and concepts.

Let's start by understanding the different **data categories** you manage in your Salesforce org:

- **Master data**: This is data that is typically part of many types of systems within an enterprise, and as such, needs to be mastered in one system:

  - Examples include customers, suppliers, products, and assets

- **Transactional data**: This is data that is created as a result of doing business with your customers, partners, and suppliers:

  - Examples include opportunities, quotes, orders, payments, cases, and activities

- **Metadata**: Your Salesforce org is made up of metadata – the configuration and customization you have done to make Salesforce work for your organization.

- **Reference data**: This data is a necessary supplement to the master data (but is *not* the same thing), as it provides context for your transactional data. It may come in the form of **metadata** or **configuration data**:

  - Examples include state and country picklist values, ISO codes, loyalty scheme tiers, customer segmentation names, and classification

- **Reporting data**: This includes any data you have created solely for the purpose of reporting.

> **Important note**
>
> Be sure to align your Salesforce reference data with your organization's enterprise architect and data architect to ensure consistency across systems.
>
> In *Chapter 6, Detailing the Scope and Design of Your Initial Release*, we introduced the concept of a **data dictionary**. Make sure you keep it up to date beyond your initial Salesforce release so that it can add value to your stakeholders—the consumers of your Salesforce data—and accelerate development efforts for your continuous improvement teams.

For each category of your Salesforce data objects, it is important for you to consider the life cycle of your Salesforce data – by data category and object. So, let's get on the same page regarding what the data life cycle is:

- **Creation**: When is the data created and who can create records?

  - If the data is system-generated, what triggers that?

- **Storage**: When is the data stored in your Salesforce org and what level of encryption is required (if any)?

- **Usage**: Who needs to access the data, for what process, and for how long?

  - For how long does your organization need and want, or is allowed, to store the information?

- **Archival**: When is it no longer value-adding to keep the data in Salesforce? What processes and tools (for example, batch jobs or enterprise tools) do you have in place to manage regular archiving (and backup process)?

> **Tip**
>
> Consider using an enterprise Salesforce backup and restore solution to ensure business continuity, as well as performance and compliance. Leading vendors include **OwnBackup** and **Odaseva**.

- **Deletion**: When should the data no longer exist in Salesforce or at all within your organization?

With these concepts out of the way, let's move on to data quality.

Governing the quality of your Salesforce data

There are two sides to **data quality**. If your data quality is good generally, this is an enabler for a delightful user and customer experience and ultimately great business outcomes. Conversely, if your data quality is generally poor, the opposite is a risk.

Let's start by looking at what **purpose** your data quality governance effort can aspire to:

- Ensuring good user and customer experience

- Enabling reliable use of data for the following ends:

  - Reporting and analytics

  - Communication

  - Automation

- Complying with legal regulations or industry standards

With the purpose of data quality governance squared away, let's look at the dimensions that define data quality:

- **Accuracy**: To what extent does your data represent the truth?

- **Age**: How often is data updated? When was it created?

- **Completeness**: To what degree are all (relevant) fields populated?

- **Consistency**: How consistent is the captured information?

- **Duplication**: What is the percentage of duplicate records per object?

- **Usage**: To what extent is the data being used by users for automation and analytics?

It's important to note that data quality isn't binary, but rather it's a spectrum ranging from very low to very high. No organization scores 100% on all the attributes of data quality all the time.

> **Tip**
> You can learn more about data quality by taking the *Data Quality* module on Trailhead: `https://trailhead.salesforce.com/content/learn/modules/data_quality`.

As part of your Salesforce CoE, you should have a forum, a governing body responsible for overseeing your efforts to drive data quality. That body is your **data governance board**.

In the following figure, you can see how our scenario company, **Packt Manufacturing Equipment (PME)**, has organized its Salesforce CoE in the continuous improvement phase of its Salesforce program:

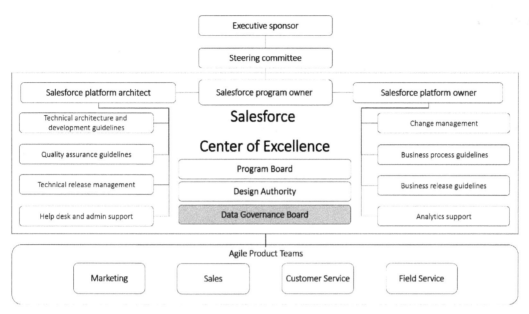

Figure 13.1 – PME's Salesforce CoE with a data governance board

PME's Salesforce data governance board sits within its CoE and consists of a data steward, domain data owners, and the product owners for the Agile product teams. As governance is a philosophy rather than a management/decision function, the charter of the board is to develop guidelines and principles to further PME's Salesforce data management and quality improvement capabilities. PME's data governance board also reviews the output of any analysis and explores initiatives to improve these capabilities.

Consider the following key activities you should perform to improve the quality of your Salesforce data – using the Lean Six Sigma **DMAIC** framework:

Figure 13.2 – Lean Six Sigma DMAIC

> **Tip**
> To learn more about the DMAIC framework, go here: `https://kanbanize.com/lean-management/six-sigma/dmaic`.

Let's go through each of the steps!

Defining

Your **data steward** should lead the definition of your organization's goals for data quality across all the domains supported by your Salesforce platform. Your data steward should not be alone in this exercise, but rather work together with **data owners** – consumers of the data and representatives from the domains of your platform.

You can operationalize and make your goals concrete by using a data quality goal template:

The [Field_Object_RecordType] to be [Quality Definition] in order for [Stakeholder] to be able to [achieve business value]

An example of a practical data goal is as follows:

The "billing address on the Account object of Record Type Bill-To" to be "verified at least annually" in order for the "finance team" to be able to "invoice with minimum risk"

Measuring

Once you have defined your data quality goals, you need to track how you are performing against them.

You can – to some extent – use free tools to track and measure your data quality. These include the following:

- *The Data Quality Analysis Dashboards* from **Salesforce Labs**, which you can find here: `https://appexchange.salesforce.com/listingDetail?listingId=a0N300000016cshEAA&tab=e`

- *ZoomInfo Field Trip – Discover Unused Fields and Analyze Data Quality* by **ZoomInfo**, which you can check out here: `https://appexchange.salesforce.com/listingDetail?listingId=a0N30000003HSXEEA4&tab=e`

To take your data quality to the next level, you may consider investing in a paid solution such as **Delpha**, which offers both automatic and manual merging of duplicate records. Check it out here: `https://appexchange.salesforce.com/listingDetail?listingId=a0N3A00000FMpuXUAT&tab=e`.

Analyzing

Once you have a good method for tracking the quality of your data, you need to analyze it to spot trends and gaps. The outcome of your analyses should be ideas or initiatives to improve your data quality.

Depending on your Salesforce CoE setup and the nature of the ideas for improvement, you can decide either to do one of the following:

- Reviewing data quality analyses regularly (for example, monthly)
- Reviewing ad hoc input for data quality improvement initiatives

Depending on the scale of your Salesforce setup and CoE, it may be the responsibility of your Salesforce administrator, a central or local data analyst, or a BI/analytics team to set up reporting and track your data quality.

Improving

Your Salesforce administrator(s) or other qualified persons should clean up the data based on the output of the analysis.

This may range from record creation and updates to improving the quality of your Salesforce data. It may also entail archiving or deleting data deemed unnecessary or non-compliant to retain, or to improve your Salesforce performance and user experience.

Controlling

Your data steward should recommend initiatives to your Salesforce (domain) product owner(s), as well as your Salesforce **Platform Owner** (**PO**). The PO can, together with your Salesforce platform architect, assess and determine which initiatives to pursue to mitigate and minimize the future risk of poor data quality – and improve it.

The initiatives may include the following:

- Process re-training to ensure process compliance and quality.

- Modifying your application and system architecture and configuration. This may include creating/updating validation rules, automation, record access changes, and making changes to user permissions.

You can consider implementing tools to automate regular data operations. Some tools to consider include the following:

- Salesforce **Scheduled Flows** is a (relatively) simple declarative tool, native to Salesforce (with volume and scheduling limitations to bear in mind)

- **ETL** tools such as **MuleSoft** and **Informatica** are entirely at the other end of the spectrum

For use cases that concern publicly available business information, such as addresses, search **AppExchange** for apps to sync the data. These solutions come at a cost, so weigh the cost of the app versus the cost (in time) for your Salesforce team to (manually) update the data.

If your Salesforce team performs the same data cleansing/updating operation regularly, it's typically a straightforward business case to invest in some form of simple automation.

When considering deleting unused fields and *validating that with your product owner and potentially end user representatives*, make sure you perform the due diligence in terms of the impact analysis before this.

> **Tip**
>
> As a best practice, delete fields in a two-step process. First, remove the field from page layouts or dynamic forms and remove field-level security access. After a month, if no stakeholders have asked where the field went or complained about missing information in dashboards, go ahead and back up any records with the field populated, and then delete the field.

Data management and governance are some of the most important activities in the continuous improvement phase of your Salesforce program. Next, we'll look at how you can get started, little by little, to improve.

Getting started with data quality governance

How well you manage your Salesforce data and its quality is not binary – it's a spectrum.

If you find yourself mostly putting out fires – doing reactive data loads to rectify poor data quality, for example – you are more toward the lower end of the maturity spectrum. Not to worry, this means you have plenty of potential to improve – which will bring great benefits to your users, customers, and your organization.

> **Tip**
>
> To understand how mature your organization is at managing its data, check out this article from *Experian*: `https://www.edq.com/blog/data-maturity-how-mature-are-you/`.

So how do you get started? Let's look at an approach to consider:

1. Establish a Salesforce **data governance board** to be anchored within your Salesforce CoE.
2. Determine its charter, allocate sufficient resources, and ensure business stakeholder alignment and involvement from the start.
3. Work with your business stakeholders to understand what data and which KPIs are most important for them (refer to *Chapter 4*, *Securing Funding and Engaging with Salesforce and Implementation Partners*, for inspiration).
4. Evaluate and determine which tools you will use to support you in analyzing and managing your Salesforce data.
5. From here on, follow the DMAIC model laid out earlier in this chapter.
6. Lastly, don't attempt to determine, analyze, and fix everything at once:

    A. Similar to your evolved Salesforce DevOps capabilities for bringing new features to your users, you should strive to **deliver consistent, incremental value** by continuously improving the quality of your data

> **Tip**
>
> To learn more about Salesforce data management and data quality, you may read the following:
>
> - *Salesforce Data Architecture and Management* by Ahsan Zafar (CTA) (`https://www.amazon.com/Salesforce-Data-Architecture-Management-effectively/dp/1801073244`)
>
> - The article *Data Management Best Practice Guide* from Salesforce here: `https://help.salesforce.com/s/articleView?id=000393223&type=1`

Next, let's turn to how you can enjoy the fruits of your investments in Salesforce and the quality of your data.

Leveraging your C360 data to maximize the return on your Salesforce investments

Let's get down to business by first assessing your organization's data leverage maturity.

There are different ways in which you can leverage the data at your disposal. If you think about how you use your Salesforce data, you will likely be able to allocate this to one of the following stages:

- **Descriptive**: *Knowing what* happened

- **Diagnostic**: *Knowing why* something happened

- **Predictive**: *Foreseeing* what is likely to happen

- **Prescriptive**: *Suggesting* actions to manage or leverage what is predicted to happen

- **Cognitive**: *Acting* based on events occurring and/or based on anticipation

> **Important note**
>
> You can't jump from descriptive to cognitive analytics. You have to progress through the stages. The critical success factors for mastering and gaining value from each level are quality data and the continuous validation of the accuracy and value that the initiative is bringing. The validation part is often missed by organizations attempting to skip steps. Validation is best done by a combination of visionaries imagining and striving for the next frontier for the business and people with intimate knowledge and context of the use case. These may be domain super-users who are most eager to rid themselves of rudimentary, repetitive tasks or who wish to improve a sales or service process.

With the theory out of the way, let's dive into how you can leverage your Salesforce data to unlock value for your users and organization.

There are two main ways to leverage your Salesforce C360 data. These are as follows:

- **Knowledge and insights:**

  - As the basis for optimized, data-driven decision making

- **Automation:**

  - Of repetitive, responsive manual tasks for your users

  - Of responsive or proactive actions

> **Important note**
>
> These two ways are not to be seen as discrete or disjointed, rather with the first being the catalyst for the latter. As your organization matures its ability to leverage your C360 data, the subtle line dissipates and will seem to fade.

Let's look at potential use cases for leveraging your Salesforce C360 data.

Next, let's explore some of the most obvious and common use cases for leveraging your Salesforce C360 data by business domain and across business domains:

- **Marketing**: Use cases include marketing operations insights and automation, including initiating or progressing customer journeys triggered by events.

- **Sales**: Use cases include sales productivity powered by Sales Cloud **Einstein** for insights on what leads are most likely to convert and which opportunities are most likely to close. For commerce, product recommendations are a great use case for Einstein.

- **Service**: There are numerous use cases for leveraging your Salesforce data to drive efficiency in your service organization. Key ones include case management productivity, such as automating case creation (web-to-case, self-service), or case assignment by combining Einstein Case Classification with **Omnichannel** skills-based routing. If you are currently providing (or considering providing) chat functionality to your customers, you should consider implementing chatbots to enable your customers to self-serve to deflect simple cases from being created in your Salesforce org. Start with menu-driven chatbots, and gradually, as you gather more data from chat, progress to intent-driven **AI chatbots**.

- **Cross-domain**: For all domains, there are Salesforce platform features you can leverage to reach higher levels of productivity. These include **Next Best Action**, **CRM Analytics Predictions**, and **Salesforce Genie***, among many, many others.

> **Tip**
>
> *Salesforce Genie is a new layer or suite of product offerings from Salesforce. You should do your own research and make any decisions based on the product information available. Learn more about the promise of Salesforce Genie at `https://www.salesforce.com/products/genie/overview/`.
>
> To learn more about AI and Salesforce, you can read *Architecting AI Solutions on Salesforce* by Lars Malmqvist (CTA) (`https://www.amazon.com/Architecting-Solutions-Salesforce-state-art/dp/1801076014/`).

In addition to analytics and automation use cases *within* your Salesforce org, let's consider use cases that integrate and automate processes with *external* systems:

- **MuleSoft RPA**: Using MuleSoft to, amongst other things, use **Optical Character Recognition (OCR)** to process received documents and transform these into system-consumable data:

  - Use cases may include receiving an order by PDF, and instead of a human manually creating the order, MuleSoft can do it for you.

- **MuleSoft Composer**: A declarative (no-code) solution enabling admins and functional consultants to set up cross-system process automation:

  - Although Composer is a declarative solution, make sure you follow the same design governance rigor as you would for any integration – require solution designs to be reviewed by your **Design Authority**.

  - Use cases include posting case updates to a Slack channel. A couple of things to consider are as follows:

    - Composer is designed for low-volume processes, not batch or bulk use cases.

    - There is no concept of sandboxes for testing, so be diligent when "turning on integrations" by connecting Composer flows to production.

- **MuleSoft Anypoint**: Anypoint is a leading, enterprise-grade middleware (ESB) and ETL tool and is part of the Salesforce portfolio of products. If you don't have middleware or ETL on your enterprise architecture roadmap, you wish to pursue IT strategies such as **API-led connectivity/API economy**, and you are already using Salesforce for CRM, you should consider MuleSoft.

Next, let's prioritize your initiatives.

Prioritizing your analytics and automation initiatives

As you identify potential initiatives for leveraging your Customer 360 data, it's useful to know what to consider when prioritizing which initiatives to pursue.

Let's cover the overall objectives and points you should consider for each type of initiative:

Initiative type	Key objective	Key points to consider to create your business case for the initiative
Analytics	Gain insights to optimize decision-making	What's the risk of not knowing (early or immediately)? For these initiatives, plotting the risks of different initiatives on a simple impact-likelihood matrix will provide you with guidance to choose the most valuable analytics initiatives to invest in. Weigh this up against the effort to deliver the insights in Salesforce.
Efficiency	Alleviate rudimentary tasks	How often (per month or year) does the task occur and how long (in minutes) does it take? Extrapolate out to calculate the potential time saved for all users who will benefit from the automation. Weigh this up against the effort to deliver the automation in Salesforce.
Revenue	Drive sales	How will faster, better, more reliable insights or automation increase your opportunity win rate and average deal size, or reduce their close time? Weigh this up (accounting for the average gross margin of your organization) against the effort to deliver the automation in Salesforce.

Table 13.1 – Business case considerations by type of initiative

Keep only **one** product backlog for prioritizing both your data analytics and automation initiatives, against your other Salesforce development initiatives.

> **Important note**
>
> The law regulating the development and use of AI is a quickly evolving legal field. Work closely with your legal team to understand the laws you need to comply with in regard to the management of your customer data and AI services.

Let's move on to the final topic of this chapter – inspiration!

Leading and inspiring the peers in your organization

When you deliver value to your stakeholders, the word will spread that your platform, your program, and your team(s) are delivering improvements for your organization.

Likely, when this starts to happen, you – and your Salesforce program and team – will have evolved the capabilities of delivering new functionality (DevOps), as well as analytics, and potentially AI services (DataOps and AIOps).

Take the praise humbly and share the knowledge and learning from your journey to get where you are; you have arrived at the spearhead of your organization, leading its elusive **digital transformation**.

Let's wrap up this chapter!

Summary

We covered a lot of topics in this chapter of the continuous improvement phase of your Salesforce program. First, you learned how to manage the life cycle of your Salesforce data and govern its quality to set your analytics and automation initiatives up for success. Next, you explored the many use cases for leveraging your Salesforce C360 data.

You have now gained insights into the methods and activities you need to continuously execute and evolve in order to **maximize business value from your investments** in Salesforce, data, analytics, and automation initiatives.

Successfully delivering on those initiatives will help you inspire your peers within your organization and allow you to take on broader and bigger initiatives.

In the next chapter, we'll discuss typical issues you may encounter in the continuous improvement phase of your Salesforce program, and how to mitigate and overcome them. We will also provide a checklist for evaluating the state of your Salesforce program.

14

Common Issues to Avoid in the Continuous Improvement Phase

In the previous two chapters, we covered the activities you will be executing in the continuous improvement phase of your Salesforce *program*.

In this chapter, we will discuss the issues most commonly faced in the ever-lasting phase of a Salesforce program. We will also cover strategies to prevent or mitigate the issues from arising altogether.

In *Chapter 5, Common Issues to Avoid in the Pre-Development Phase*, you were introduced to **root cause analysis**, and we'll continue using that method to dissect the common issues faced.

At the end of this chapter, you will be presented with a set of **checklists** that you can refer to at regular intervals to evaluate the state of your Salesforce program in the continuous improvement phase.

This chapter will cover the following main topics:

- Common issues in the continuous improvement phase and their root causes
- Evaluating the state of your Salesforce program

Let's go!

Common issues in the continuous improvement phase and their root causes

Let's get started by looking at some of the most common issues for being unable to provide a stable Salesforce platform while continuously delivering incremental business value in the continuous improvement phase – illustrated with a fishbone diagram:

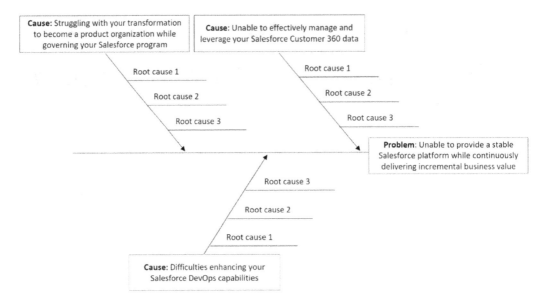

Figure 14.1 – Fishbone diagram of common issues in the continuous improvement phase

Next, we'll discuss the possible root causes and strategies for mitigating and preventing *each* of the common issues.

Struggling with your transformation to become a product organization while governing your Salesforce program

In *Table 14.1*, you'll find possible root causes for why you may be struggling with your transformation efforts to become a product organization while at the same time governing your Salesforce program – along with strategies for mitigation if you are already facing that issue:

Possible root cause	Strategies for mitigation and prevention
Senior management is not involved or hasn't bought into the concept and value of transforming to become a *product organization*, and therefore doesn't see the need to transform.	The concepts, principles, and approaches of a product organization are key to unlocking the value of your Salesforce program (refer to the *Transforming your organization to become a product organization* section in *Chapter 12, Evolving Your Salesforce Org and DevOps Capabilities*). If your original pitch of the concept to senior management didn't end with a standing applause, don't be discouraged – it rarely happens. Organizational transformation is a large change and should be treated as such. Rather than an all-or-nothing approach, here is a more palatable course of action for you to consider: Get a mandate to experiment with the product organization concept for your Salesforce program only, not across the entire organization. If it's not feasible for your entire Salesforce program, try to get buy-in from one senior manager/department to experiment with the approach for the business area. Incorporate as many of the practices of a product organization as you practically can within your Salesforce program – within the sphere of your control. As you begin to see positive results, share them with your peers and senior management – with the intent of inspiration. Continuously use the language of product organizations when interacting with your peers and senior management. This will gradually diffuse the concept and your stakeholders will slowly adopt it via osmosis.

Your organization continues *funding* discrete projects rather than your Salesforce program as a whole, including consistent agile product teams.	This is tightly related to the previous root cause. Funding agile product teams that continuously deliver value is fundamental to being a product organization.
	If your organization has traditionally been using waterfall project methodologies with SoWs based on comprehensive BRDs, the shift to autonomous agile product teams will seem radical – and not likely to be something that will be instantly supported by senior management.
	What you need to do is remain determined that the goal remains the same while proposing a compromise. It could be to implement a mid-term planning mechanism, such as SAFe program increments. Pure agilists will be quick to call SAFe a scaled waterfall framework. While that may be (partially) true, even a small change in "how things are done" is better than no change as it will get you started on the journey.
	In this way, you are leveraging SAFe as a component in your transition architecture for your delivery transformation journey.
	As you progress, highlight the benefits of this new way of working for your organization. Instill and encourage the mentality of kaizen (the Japanese philosophy of change for the better via continuous improvement) and use sprint retrospectives to understand what changes you can make to improve your way of working.

Your Salesforce program and platform *vision* and *strategy* are disjointed or misaligned from your organization's *vision* and *strategy*. As a result, you struggle to set goals for your Salesforce platform and its products. A symptom of this root cause is that your Salesforce teams may be very productive – churning out user stories at an impressive rate – yet your Salesforce program is *not being perceived as delivering value* to your organization.	Alignment of your Salesforce program is fundamental to success in the continuous improvement phase. So, what can you do to get aligned? Refer to the original business case that you prepared before you start developing your initial release. Read *Chapter 4, Securing Funding and Engaging with Salesforce and Implementation Partners,* to revisit the components of your business case, along with *Chapter 1, Creating a Vision for Your Salesforce Project,* to understand and (re)define the vision for your Salesforce program. You should be crystal clear on who the customers (users) of your Salesforce products are – who your Salesforce platform serves. You can use the **Jobs-To-Be-Done** (**JTBD**) framework, wherein you evolve their persona descriptions across the following dimensions: • **JTBD**: Functional and emotional jobs your personas (users) have to do. • **Pains**: The risks and obstacles those personas are facing. Consider how your product may help alleviate these pains. • **Gains**: The desired outcomes are benefits your personas are aiming for. Make sure *all* members of your agile product teams, Salesforce CoE, and wider stakeholders have access to and understand the persona descriptions. Rich persona models are great conversation starters – especially in combination with swim lanes in business process maps – to understand who, what, and why your Salesforce products are or should be a certain way.

You requested *funding* for the following core Salesforce CoE roles: • Salesforce program owner • Salesforce platform architect • Salesforce platform owner (chief product owner) • Salesforce product owners for each agile product team But you didn't manage to get the full budget approved. You may only have received funding to staff 2 or 3 of those roles.	Without proper funding, you can't hire people to oversee, manage, and continuously deliver improvements for your users, customers, and organization. Funding needs to cover both your Salesforce CoE as well as the agile product teams and potential new projects. To be pragmatic and make the most of your funding, you have the following options, each with downsides to consider: • Combine the platform owner and program owner roles. This is often a preferred compromise. Be aware of the resulting reduced focus you would otherwise have had if there were different persons responsible for different aspects within your Salesforce program. • Let some of the roles have less than full allocation to your Salesforce program. This will also result in a reduced focus as the roles will have other competing priorities. Regardless of your funding, make the most of it. As you progress your Salesforce program and deliver value, you can shape a business case to reach new heights.
The *ownership* and decision-making mandate of your Salesforce product roadmap remain at senior leadership – or a collective of division/region heads. In addition, your product *roadmap* is a *time-based* representation of releases.	If prioritization and decision-making are (purely) top-down rather than bottom-up, you risk building features that aren't validated to deliver value to your customers (users). Another consequence is the difficulty updating your roadmap regularly. What would be better – and what you need to try to convince your senior management of – is that your product roadmap should represent the intention for product *evolution*. Rather than top-down prioritization, you should aim for empowered product owners who engage with end users to create a more impactful roadmap, which can be updated regularly and – importantly – communicated to your stakeholders.

Your *operating model* – and herein your *partner model* – is not supportive or aligned with your Salesforce program ambitions. Indications of this root cause are that your organization, including your Salesforce CoE, lacks the *competencies* to fulfill the vision of a product organization.	If you have great ambitions for your Salesforce roadmap – meaning you wish to deliver many features that represent great business value for your users – but not the operating model to support it, it's just wishful thinking. Consider who you intend to be part of the continuous delivery of your roadmap. Refer to the *Adjusting your partner model in the continuous improvement phase* section of *Chapter 12, Evolving Your Salesforce Org and DevOps Capabilities*, to reflect on the future operating and partner model for your Salesforce program.
You are *over-investing* in *product development capabilities* (how to build) at the expense of *product discovery* (what to build and why).	Without mature means and experienced people interacting with your users to understand their jobs to be done, as well as their pains and gains, you will struggle to empathize with them and – as consequence – deliver sub-optimal solutions. Avoid confusing Agile (a mindset and an umbrella of methodologies for *how* to collaborate to design, build, and test something) for being a method to determine *what* to build and *why*. Yes, the *Agile Manifesto* states user interactions as being key but does not further describe how to do so. Here are some ways you can enhance your product discovery capabilities: • Make sure customer experience designers are part of your agile product teams. • Facilitate regular user feedback forums with different user groups (personas) where you – in addition to getting feedback on existing features and usage – invite them to share their views on prioritization of the product backlog. Take their feedback into consideration when planning your roadmap. • Keep channels open for user feedback – and respond to it with gratitude.

Your agile development teams and their product owners are stuck in the *build trap*, believing all problems and feature requests can only be solved with a technical solution.	Technology is great and many problems can be solved relatively quickly on a platform such as Salesforce. But the truth is, not all problems require technical solutions. Before considering the solution type, do research and analysis to understand what needs to be achieved and why. Only then can you determine what the optimal solution type may be. It may be a technical solution, but often, it may be training, communication, or a third solution type.
You have (with good intention) implemented SAFe or another framework to scale agile *verbatim*, believing it will automatically make your organization agile. Instead, you are drowning in overhead due to planning and ceremonies (meetings).	A framework should work for *your* benefit, not the other way around. You should – mindfully – adapt the parts of a framework that make sense in your organizational context, and make it your own. If your Salesforce program is your organization's first encounter with agile, consider engaging with a partner to guide you on your transformation journey to becoming agile at scale.

Table 14.1 – Possible root causes for struggling with your transformation to become a product organization while governing your Salesforce program

Let's jump to the next common issue in the continuous improvement phase.

Difficulties enhancing your Salesforce DevOps capabilities

In *Table 14.2*, you'll find possible root causes for why you may have difficulties with enhancing your Salesforce DevOps capabilities – along with strategies for mitigation if you are already facing that issue:

Possible root cause	Strategies for mitigation and prevention
You are *not evolving* your Salesforce *DevOps capabilities*.	DevOps itself is an evolving field. The same is true for the Salesforce platform and related technologies. If you aren't leveraging these improvements, you're not able to get the most out of your investment in the platform – for the benefit of your users and organization. This means DevOps isn't something you do, it's a capability you should aim to continuously improve. It may seem overwhelming to comprehend where to begin, so consider the following steps as a starting point: 1. Map out your current DevOps process. 2. Identify areas of improvement that would bring the most value – improve the DevOps metrics (refer to DORA). The areas could be any of the ones mentioned in the *Enhancing your Salesforce DevOps capabilities* section of *Chapter 12, Evolving Your Salesforce Org and DevOps Capabilities*. 3. Understand it's a journey, so implement improvements *incrementally* – adopting DevOps doesn't mean you have to make major changes all at once. Rather, taking just one small step can bring noticeable rewards. 4. Assess the benefits realized by your improvements and continue onwards.
You struggle to achieve *adoption* of newly released features.	The issue may not be the features themselves, but the way they are released. What you need to do is remember change management, communication, and training when releasing new features. Refer to the *Planning, communicating, and managing change* section in *Chapter 10, Communicating, Training, and Supporting to Drive Adoption*, for inspiration regarding what to do when releasing new features. Be mindful of and calibrate your efforts in the continuous improvement phase – minding the principle of proportionality to the change impact your new features represents. Make sure you create reports to track and *know* how your users are adopting your new features. Compliment usage reports with user interviews to understand adoption and the potential for improvements.

Your production *deployment duration* is increasing and/ or your *lead time* from the new request to release is high or increasing. As a result, you struggle to meet your *sprint or PI goals*.	If you're facing this as a root cause, many consequences make taking action to remedy it imperative: • *Unhappy* Salesforce team members (the stewards of your platform) • Loss of organizational trust in your Salesforce program, and ultimately loss of funding • Long wait for new features • Wasted resource (time) by your team members that could have been spent developing new features • High risk of deployment failure • High risk of lower adoption Often, the issue is related to not evolving your DevOps capabilities. Consider the following approach to reduce deployment and lead time: • Recall and aim for the principles of DevOps – smaller, more frequent releases reduces the risk and complexity of the deployment and change in management efforts • Aim to progress from full deployments to partial, smaller deployments • Use the theory of constraints to reduce bottlenecks in your DevOps process • Train and upskill team members on your CI/CD tools (and put in place checks for QA), or outsource technical deployments management • Consider automating manual deployment tasks (the non-deployable components) by setting up advanced scripting as part of your CI/CD pipeline If the transition to smaller, frequent releases is not currently an option for your Salesforce program – for any number of reasons – you need to increase the vigilance of keeping the deployment plan (runbook) updated.

You struggle to *translate* feature requests – large and small – into appropriate product backlog items or projects, and as a result, you struggle to decide on the appropriate delivery approach for requests.	If you are not able to effectively triage requests, and size them up in your initial qualification of the requests, you risk not being able to progress them. Worse yet, you might mistake a request for being a big project – and spend (wasting) many people's time deciding how best to implement it – for what is a small request with a simple fix. In the opposite situation, expectations of a fast delivery will soon turn to despair and delay.

While you should embrace the more pure agile delivery methodology for product *evolution* of existing capabilities supported by your Salesforce solution, it isn't always the best approach for all initiatives.

Involve a combination of your product owner, business analyst, customer experience designer, and solution architect to assess requests and translate them into appropriate product backlog items or projects.

Decide on a framework for organizing your product backlog item hierarchy. One example is described in *Table 12.1 - Salesforce product backlog item types*.

Refer to the *Understanding delivery methodologies* section of *Chapter 3, Determining How to Deliver Your Salesforce Project*.

If you evaluate a request to constitute a discrete project, follow *Parts 1-3* of this book. If not, refer to *Chapter 12, Evolving Your Salesforce Org and DevOps Capabilities;* your agile product team should deliver according to your DevOps process. |
| Difficulties managing multiple *concurrent* agile product teams and long-running projects.

This could be due to still being in the process of *rolling out* your initial (MVP) release to more geographies. | If you're encountering this issue, you likely face difficulties *prioritizing* or allocating resources to continuously improve your Salesforce products, as well as potential team burnout due to too many commitments running at the same time and risk to delivery.

What is key to remember is the ancient art of managing expectations based on realistic assumptions of what is possible with the resources at hand. If more is expected than currently possible, either reduce expectations (move roll-out timelines) or add more resources – either for roll-out or the continuous improvement.

A prerequisite for concurrent streams and long-running projects is a sophisticated CI/CD pipeline and tooling. Be sure to consult a technical architect and DevOps expert to guide you in the best course of action for your Salesforce program to realize your ambitions. |

Your product roadmap does not include technical enablers, also known as an *architectural runway*.	This often happens when there is no involvement of a Salesforce architect (platform, technical, solution) in updating a Salesforce product roadmap or when assessing new feature requests. Without a technical runway, delivery risk and potential technical debt accumulate – impeding your agility in the long term. You need to establish a practice of involving a Salesforce architect in your process for evaluating new requests. Next, you need to allocate appropriate time and architect resources for upfront architecture and design for medium and large backlog items – and for any project-type initiative.
You have no structured approach for continuously dealing with *technical debt*.	This is often closely related to a lack of an architectural runway. The symptoms of unaddressed and mismanaged technical debt appear due to your team members drowning in increasing amounts of required maintenance and intermediate bug fixes. At the same time, your stakeholders may have their patience tested while waiting for new features to be developed and released. As discussed in the *Managing technical debt* section of *Chapter 12, Evolving Your Salesforce Org and DevOps Capabilities*, this mainly arises from: • Changes made to your Salesforce platform based on poorly designed solutions that require a high degree of maintenance, which outweigh the benefits of the solution itself • Poorly governed solutions that have been extended beyond what they were intended and suitable for What you can do about it is a combination of *prevention* and *alleviation*: • Your PO should initially review and qualify bugs and new feature requests. If it's not a valid bug (a broken feature) or a new request aligned with your product goals, no solution for it should be implemented. • Implement a design authority to review proposed solution designs before development can commence. • Ensure development guidelines are in place, communicated, understood, and used. They should also be continuously updated and refined.

	• Require agile product team members to explicitly state the required maintenance effort for their proposed solutions on the backlog item for review. Challenge solutions and the corresponding business requirements that require maintenance – or where the section is missing altogether. • Allocate 15%-25% of the PI or sprint capacity for continuous refactoring. This could either be by including refactoring backlog items in sprints, or by having a dedicated "refactoring" sprint at the end of each program increment. While the latter is less DevOps-friendly, it is sometimes easier to grasp, manage, and plan (and therefore it happens) – making it an attractive alternative to the optimal DevOps approach. You can read this article by *Ian Gotts* to further understand technical debt: `https://medium.com/salesforce-architects/how-architects-can-help-reduce-technical-debt-c0814bf8eaa`.
Documentation and *training materials* are poorly maintained post-initial go-live.	This is an issue for obvious reasons; without documentation, new members will struggle to understand what your Salesforce platform does and why. It's also an issue for existing team members to know what the baseline is for any changes. Training material – in whatever format – is critical for user adoption of existing and new users. Keeping documentation and training material in decent shape relies on your people, process, and tools. Incorporate updating documentation (for example, business process maps, data models, data dictionaries, training documents, and more) in your definition of done for user stories. For documentation, consider collaboration tools such as **LucidChart** that have pre-built Salesforce templates to start from. Consider leveraging org analysis tools such as **Elements Cloud**, **Salto**, and **Metazoa**.

You are overwhelmed with *bugs* and, after addressing a bug, you later encounter the same or a similar bug.	Your bug reports may range from reports of poor *performance* and broken or incoherent *automations* to *cluttered UI* and cumbersome *navigation*. Bug fixing is cumbersome and time-consuming. While for some it may be mentally stimulating to troubleshoot and identify the root cause of a bug, in reality, it's a waste of time. Fixing bugs is not a waste, but the time could have been better spent had it been prevented. What should you do if you experience a large volume of bugs? 1. *Fix*: Spend a proportional amount of time investigating the root cause of a bug by using the **5 Whys method** to uncover and fix the underlying human causes of seemingly technical issues. Check out Eric Ries' video on the method here: `https://hbr.org/2012/02/the-5-whys.html`. 2. *Prevent*: Implement a design authority to review proposed solution designs *before* development can commence. Also, dedicate time to improving development standards, guidelines, your testing framework, and the developer onboarding process.

Table 14.2 – Possible root causes for having difficulties enhancing your Salesforce DevOps capabilities

Let's jump to the next common issue in the continuous improvement phase.

Unable to effectively manage and leverage your Salesforce Customer 360 data

In *Table 14.3*, you'll find possible root causes for why you may be unable to effectively manage and leverage your C360 data – along with strategies for mitigation if you are already facing those issues:

Possible root cause	Strategies for mitigation and prevention
	High-quality data is the foundation for anything you want to use Salesforce for.
	Follow the DMAIC approach described in the *Governing the quality of your Salesforce data* section of *Chapter 13, Managing Your Salesforce Data to Harvest the Fruits of Customer 360*.
	You may take the following approach to try to identify the underlying causes of reports of poor data quality:

You receive many reports of *poor data quality* and you struggle to identify the underlying causes.	1. Determine what the *data category* is for the object(s) that are reported to have poor data quality.
	
	2. Understand what *dimension* of data quality is reported to be poor.
	
	3. *Assess* how poor the quality of the data is. If it is indeed very poor and has a significant business impact, continue.
	
	4. Understand the *business process* the object is related to.
	
	5. *Analyze* to find trends and patterns that could lead you to the root cause. Places to look include the data life cycle (is the quality issue arising when the record is created, processed, or used?), record type, creator's or owner's role, sales organization, territory, category, stage/status, data access rights, and permissions:
	
	    I. Check both the reporters and those creating and editing the data – sometimes, data quality isn't the issue, it's data access or permissions.
	
	    II. For the core *hero* transactional objects (Lead, Opportunity, and Case); *age* is a key data quality dimension and general KPI to track.

You encounter difficulties establishing a data governance body and *anchoring ownership* within your organization.	You need to ensure *data owners* are allocated and own the data for their business domain. If this is not the case, raise the issue in your Salesforce *data governance board* to make resources available for data ownership.

You struggle to *prioritize* your data quality improvement efforts.	In the *Define* section of DMAIC in *Chapter 13, Managing Your Salesforce Data to Harvest the Fruits of Customer 360*, we looked at the concept of *data quality goals*. If you have already set these for your organization, that is the data you should aim to improve the quality of. If you have yet to establish your data quality goals, the following types of data should be considered for prioritization: • Personal data (Contacts and Leads), which is used for communication, so if the data quality is poor, you risk poor customer experiences, satisfaction, and ultimately customer lifetime value. • Master data with implications and use across your org. This may include Accounts, Products, and Assets. • Data holding any type of financial information (quotes, orders, invoices, and payments). • Integrated data – if you're sending data to other systems inside or outside your enterprise, make sure it's of high quality. The same goes for receiving data – make sure you can trust its quality, or validate it upon entry. • Reference data. Other types of data may be extremely important to your organization, so be sure to make an assessment when setting your data quality goals. Data being of high quality all the time is the ambition. But when you have to prioritize your efforts to improve it, consider the timing of when the data needs to be accurate and complete. A Lead, for example, may not need to be 100% complete on creation – just enough to be able to contact it and confirm information and interest. At Lead conversion though, the quality does need to be high.
Data model changes are introduced with little or no thought regarding their *implications* for integrations, automations, or user experience – not to mention your analytics services.	Ensure any user story containing data model changes is presented and reviewed by your *design authority before* commencing development. Communicate this guideline to all team members – internal as well as external.

You struggle to find *inspiration* to leverage your Customer 360 data.	Refer to the *Leveraging your C360 data to maximize the return on your Salesforce investments* section of *Chapter 13, Managing Your Salesforce Data to Harvest the Fruits of Customer 360.* In addition, you can find inspiration in a multitude of ways, including: • Engaging with the Salesforce ecosystem – the *Trailblazer* community by participating in group events: `https://trailblazercommunitygroups.com/` • Reading books, such as *Architecting AI Solutions*, by Lars Malmqvist: `https://www.amazon.com/Architecting-Solutions-Salesforce-state-art/dp/1801076014/ref=tmm_pap_swatch_0?_encoding=UTF8&qid=&sr=` • Visiting the Salesforce AI Research website for further inspiration at `https://www.salesforceairesearch.com/`
You have initiated data analytics, automation, or AI initiatives *before* clearly *understanding the use case*, its context, and the consumer/beneficiaries (users, customer) of the initiatives.	If you don't fully understand the context of a data analytics initiative, you will not be able to develop a solution that hits the nail on its head. Best case, the use case is generic, and your solution follows general best practices for the pattern. In the worst case, your solution misses the target, leaving a trail of despair and technical debt behind. It cannot be said enough, but involve your intended users when designing solutions – the same applies to data analytics, automation, and AI initiatives.
You have initiated data analytics, automation, or AI initiatives *without ensuring proper data quality*.	This is a recipe for failure that causes poor experiences, errors, low satisfaction, and potentially lost revenue. You shouldn't develop data analytics, automation, or AI initiatives without putting in place guardrails for the quality of the data to be leveraged.

You struggle to *govern*, *maintain*, and *improve* your data analytics and AI deployments, such as chatbots or Einstein predictions.	AI solutions require continuous monitoring and improvements to deliver maximum benefit. Continuous improvements are also required to keep up with your business, customers, and users. As such, you need to account for ongoing maintenance effort of roughly 10%-20% of your initial effort in setting it up.
You struggle to see *progress* in efforts to leverage your Customer 360 data.	The principle of aggregation of marginal gains also applies to data analytics, automation, and AI initiatives. Each initiative you successfully deliver – each feature, each data quality enhancement, and each analytics capability – all contribute to and amplify the success and value you and your Salesforce colleagues bring to your organization. Remember, it's a journey – keep pushing onwards and upwards and the results will follow.

Table 14.3 – Possible root causes for being unable to effectively manage and leverage your Salesforce Customer 360 data

Next, let's evaluate the state of your Salesforce program in the continuous improvement phase.

Evaluating the state of your Salesforce program

We have covered a lot of topics and activities in the continuous improvement phase of your Salesforce program. If you get stuck, or you are not pleased with your progress, the following section will support you in evaluating the state of your Salesforce program.

For each item you cannot check, go back to the related chapters and complete the activities to mature your capabilities to maximize the return on your investments.

Checklist for your Salesforce program in the continuous improvement phase

At the beginning of the continuous improvement phase of your Salesforce program – after the first go-live with Salesforce – you should go through the following steps:

- Make sure your Salesforce program and platform vision align with your organization's vision and strategy

- Update your Salesforce CoE with core roles

- Understand the concept of product organizations

- Understand who the customers (users) of your Salesforce platform and products are

- Understand what your Salesforce products are

- Recruit or dedicate Salesforce product owners and empower them to truly own their products

- Set goals for your Salesforce platform and products – aligned with your platform vision

- Transform from funding discrete projects (statements of work) to funding your Salesforce program consisting of agile product teams

- Update your operating and partner model

- Update your Salesforce roadmap

- Establish cross-functional agile product teams

- Assess your agile maturity and DevOps capabilities – and initiate actions to evolve

Throughout the continuous improvement phase, you should *continuously* do the following:

- Enhance your understanding of product organizations and drive activities to drive your organization's transformation to become one

- Evolve your persona descriptions by deepening your insights into your Salesforce users' jobs-to-be-done and pains and gains, and regularly facilitating user feedback forums

- Monitor and understand your users' adoption and usage of new and existing features

- Govern the health of your Salesforce platform by:

  - Facilitating regular design authority sessions for reviewing new proposed solutions

  - Performing regular health checks

  - Keeping documentation up to date

  - Defining and communicating development standards and guidelines

- Move toward smaller, more frequent releases
- Regularly evaluate your Salesforce DevOps capabilities and drive actions to enhance them
- Govern your Salesforce data quality efforts by running a data governance board
- Improve your Salesforce data quality by using the DMAIC framework
- Assess your data maturity by determining the most common nature of your data analytics initiatives (descriptive-diagnostic-predictive-prescriptive-cognitive)
- Implement new and continuously improve your data analytics, automation, and AI initiatives
- Enhance your data analytics, automation, and AI capabilities to leverage your Customer 360 data
- Govern, monitor, maintain, and improve your data analytics, automation, and AI initiatives
- Inspire the peers and leadership in your organization
- Inspire and engage with the Salesforce **Trailblazer** community
- Share your knowledge and experiences

Let's wrap up this final chapter.

Summary

In this chapter, you have become familiar with some of the most common issues faced in the continuous improvement phase of Salesforce *programs*. If you face one or more of the issues mentioned in your own Salesforce program, you now know how to mitigate or prevent them from arising altogether.

You have also assessed the state of your Salesforce program by going through the phase checklists so that you know where to focus your efforts to improve.

With your Salesforce program and platform evolving, I leave you with a final note:

> *We all face challenges, risks, and issues in our work and personal lives. It's how we deal with them that shapes us and our organizations.*

> *I wrote this book because I wanted to share what I've learned while helping organizations implement Salesforce and get the most out of their investments. My purpose was to offer insights into what I've seen work and not work, so that you could learn from them, and get the most out of your time working with Salesforce.*

> *If you have found parts of this book useful or if there are parts you have a different perspective on, I'd love to hear about them and learn from them. I invite you to please share your points of view so that we can create #BetterSalesforceDelivery.*

Index

www.packtpub.com

Subscribe to our online digital library for full access to over 7,000 books and videos, as well as industry leading tools to help you plan your personal development and advance your career. For more information, please visit our website.

Why subscribe?

- Spend less time learning and more time coding with practical eBooks and Videos from over 4,000 industry professionals

- Improve your learning with Skill Plans built especially for you

- Get a free eBook or video every month

- Fully searchable for easy access to vital information

- Copy and paste, print, and bookmark content

Did you know that Packt offers eBook versions of every book published, with PDF and ePub files available? You can upgrade to the eBook version at packtpub.com and as a print book customer, you are entitled to a discount on the eBook copy. Get in touch with us at customercare@packtpub.com for more details.

At www.packtpub.com, you can also read a collection of free technical articles, sign up for a range of free newsletters, and receive exclusive discounts and offers on Packt books and eBooks.

Other Books You May Enjoy

If you enjoyed this book, you may be interested in these other books by Packt:

The Salesforce Business Analyst Handbook

Srini Munagavalasa

ISBN: 9781801813426

- Create a roadmap to deliver a set of high-level requirements
- Prioritize requirements according to their business value
- Identify opportunities for improvement in process flows
- Communicate your solution design via conference room pilots
- Construct a requirements traceability matrix
- Conduct user acceptance tests and system integration tests
- Develop training artifacts so your customers can easily use your system
- Implement a post-production support model to retain your customers

Salesforce Anti-Patterns

Lars Malmqvist

ISBN: 9781803241937

- Create a balanced system architecture by identifying common mistakes around on- and off-platform functionality and interfaces
- Avoid security problems that arise from anti-patterns on the Salesforce platform
- Spot common data architecture issues and discover intuitive ways to address them
- Avoid the dual traps of over- and under-customization in your solution architecture
- Explore common errors made in deployment setups, test strategy, and architecture governance
- Understand why bad communication patterns are so overlooked in architecture

Becoming a Salesforce Certified Technical Architect

Tameem Bahri

ISBN: 9781800568754

- Explore data lifecycle management and apply it effectively in the Salesforce ecosystem
- Design appropriate enterprise integration interfaces to build your connected solution
- Understand the essential concepts of identity and access management
- Develop scalable Salesforce data and system architecture
- Design the project environment and release strategy for your solution
- Articulate the benefits, limitations, and design considerations relating to your solution
- Discover tips, tricks, and strategies to prepare for the Salesforce CTA review board exam

Salesforce B2C Solution Architect's Handbook

Mike King

ISBN: 9781801817035

- Explore key Customer 360 products and their integration options
- Choose the optimum integration architecture to unify data and experiences
- Architect a single view of the customer to support service, marketing, and commerce
- Plan for critical requirements, design decisions, and implementation sequences to avoid sub-optimal solutions
- Integrate Customer 360 solutions into a single-source-of-truth solution such as a master data model
- Support business needs that require functionality from more than one component by orchestrating data and user flows

Packt is searching for authors like you

If you're interested in becoming an author for Packt, please visit `authors.packtpub.com` and apply today. We have worked with thousands of developers and tech professionals, just like you, to help them share their insight with the global tech community. You can make a general application, apply for a specific hot topic that we are recruiting an author for, or submit your own idea.

Share Your Thoughts

Now you've finished *Salesforce End-to-End Implementation Handbook*, we'd love to hear your thoughts! Scan the QR code below to go straight to the Amazon review page for this book and share your feedback or leave a review on the site that you purchased it from.

`https://packt.link/r/1-804-61322-3`

Your review is important to us and the tech community and will help us make sure we're delivering excellent quality content.

Download a free PDF copy of this book

Thanks for purchasing this book!

Do you like to read on the go but are unable to carry your print books everywhere?

Is your eBook purchase not compatible with the device of your choice?

Don't worry, now with every Packt book you get a DRM-free PDF version of that book at no cost.

Read anywhere, any place, on any device. Search, copy, and paste code from your favorite technical books directly into your application.

The perks don't stop there, you can get exclusive access to discounts, newsletters, and great free content in your inbox daily.

Follow these simple steps to get the benefits:

1. Scan the QR code or visit the link below

https://packt.link/free-ebook/978-1-80461-322-1

2. Submit your proof of purchase
3. That's it! We'll send your free PDF and other benefits to your email directly